SpringerBriefs in Earth Sciences

More information about this series at http://www.springer.com/series/8897

Rodolfo Guzzi

Data Assimilation: Mathematical Concepts and Instructive Examples

Springer

Rodolfo Guzzi
System Biology Group
University La Sapienza
Rome
Italy

ISSN 2191-5369 ISSN 2191-5377 (electronic)
SpringerBriefs in Earth Sciences
ISBN 978-3-319-22409-1 ISBN 978-3-319-22410-7 (eBook)
DOI 10.1007/978-3-319-22410-7

Library of Congress Control Number: 2015948730

Springer Cham Heidelberg New York Dordrecht London

Printed on acid-free paper

Springer International Publishing AG Switzerland is part of Springer Science+Business Media
(www.springer.com)

Preface

Data Assimilation is a set of mathematical techniques allowing us to use all the information available within a time frame. This includes observational data, any a priori information we may have and a deterministic or stochastic model describing our system, and encapsulating our theoretical understanding. The mathematical basis is the estimation theory or theory of the inverse problem that is an organized set of mathematical techniques for obtaining useful information about the physical world on the basis of observation.

In a conventional problem one would use a set of known prior parameters to predict the state of the physical system. This approach is usually called a "forward problem." In the "inverse problem" one attempts to use available observation of the state of the system to estimate poorly known parameters of the state itself. In both these cases Data Assimilation can be treated as a Bayesian system. The Bayesian theorem or the law of inverse probability allows us to combine a priori information about the parameters with the information contained into observations to guide the statistical inference process.

The reason why the Data Assimilation is so effective is that it seeks to produce an analysis that fits a set of observations taken over a time frame (not just the observations made at one instant in time). This is subject to the strong constraint that the evolution of the analyzed quantities is governed by a deterministic model describing the given observation.

Due to its flexibility Data Assimilation has been applied to several fields, from the numerical weather prediction, where it was first developed, to planetary climate analysis up to the evolution of biological cells.

One begins with a forecast model, often called background. In order to make useful predictions the background must be updated frequently with noisy and sparse measurements. This procedure updates the background in light of the new observation to produce an analysis, which, under suitable assumptions is the maximum estimate of the model state vector. Later, the model is restarted from analysis and produces a new background forecast.

Data Assimilation and model forecasts can be combined into an observing system simulation experiment in order to quantify the effect of changes in the observation accuracy, type, location, and frequency on the accuracy of the numerical forecast.

The method is prone to severe limitations, because eventually the forecasted parameters diverge from the true values. The optimal procedures would combine the most exhaustive theoretical and observational knowledge taking into account the errors of observation and those due to the model.

The aim of this book is to give an in-depth and complete overview in the frame of the SpringerBrief short book concept, to construct Data Assimilation in different and emerging fields from environment to biology, as outlined in the Applications chapter.

The book serves both teachers and college students and other interested parties providing the algorithms and formulas to manage the Data Assimilation wherever a dynamic system is present.

In writing this book I have followed the history of the data assimilation evolution. The first chapter gives a wide overview of the data assimilation steps starting from Gauss' first methods to the most recent as those developed under the Monte Carlo methods. The second chapter treats the representation of the physical system as an ontological basis of the problem. The third chapter deals with the classical Kalman filter, while the fourth chapter deals with the advanced methods based on recursive Bayesian Estimation. A special chapter, the fifth, deals with the possible applications, from the first Lorenz model, passing through the biology and medicine up to planetary assimilation, mainly on Mars. The earthquake application, where the model is highly nonlinear and the error is non-Gaussian, is also reported in order to offer another point of view to the reader.

The book endeavors to give a concise contribution to understanding the data assimilation and related methodologies. The mathematical concepts and related algorithms are fully presented, especially for those facing this theme for the first time.

Finally, I want to express my special thanks to Judith De Paul and George C. Kieffer to whom I am indebted for their linguistic tips and the effective corrections, and Luca Montabone for his suggestions on the chapter of applications, mainly on Mars data assimilation.

Contents

Chapter 1
Introduction Through Historical Perspective

Abstract The first approach to the data assimilation methods started with the applications of the Newtonian equations to the motion of the planets made by Laplace and Gauss. However, only with the introduction of a probabilistic view, the dynamical equations were able to make a real forecast. The first numerical weather prediction (NWP) models, based on determinism, soon became probabilistic, mainly with the contribute of the group leaded by Von Neumann at Princeton's Institute for Advanced Study. This chapter contains the history of the evolution of the data assimilation methods, from the early determinism algorithms to the most recent probabilistic methods as the ensemble forecasting. The fundamental work by Edward Lorenz and by the group of Los Alamos, which developed the Monte Carlo method, are also reported.

1.1 From Gauss to Kolmogorov

Forecasts from models provide useful information to help predict the future state of a system. However, inevitably predictions will diverge from reality as time progresses, that is an inescapable property of dynamical systems. Data assimilation (DA) is a formal approach used to help correct forecasts by introducing information observed from the environment.

The first Dynamic Data Assimilation, which we intend to describe in this book, which is that requires the existence of data and equations or dynamic models, was made to calculate the comet's orbit, the moving bodies per excellence. This is a classical problem of two bodies where one seeks to determine the position and the components of the body at a particular time, *(Principia book III prop XLI).*

The first method of finding the orbit of a comet moving in a parabola was tried by Newton [1], using three observations. He, however, wrote, "This being a problem of very great difficulty, I tried many methods of resolving it".

When the observations are taken from the surface of Earth, the apparent position, in space, of the comet is non-given and therefore the velocity components are not determined. This fact implies that the observer has to make more observations, at different times. During the interval of time, Earth and comet have moved, so the

R. Guzzi, *Data Assimilation: Mathematical Concepts and Instructive Examples*,
SpringerBriefs in Earth Sciences, DOI 10.1007/978-3-319-22410-7_1

problem of finding the elements of a comet orbit becomes a difficult problem, as Newton wrote.

In principle, since the orbit is defined by six elements and a simple, complete observation gives us the angular coordinates of the body, three observations are enough. Being the reference system given by the neighboring fixed star, the coordinates are determined in right ascension α and declination δ. Supposing the observations are made in times t_1, t_2 and t_3 the six equations of the corresponding right ascension and declination are functions of the elements of the orbit and of the dates of observations and are represented by:

$$\begin{cases} \alpha_1 = \Psi(\Omega, i, \omega, a, e, T; t_1) \\ \alpha_2 = \Psi(\Omega, i, \omega, a, e, T; t_2) \\ \alpha_3 = \Psi(\Omega, i, \omega, a, e, T; t_3) \\ \delta_1 = \Phi(\Omega, i, \omega, a, e, T; t_1) \\ \delta_2 = \Phi(\Omega, i, \omega, a, e, T; t_1) \\ \delta_3 = \Phi(\Omega, i, \omega, a, e, T; t_1), \end{cases} \tag{1.1}$$

where Ω is the longitude of ascending node, i is the inclination to plane of the ecliptic, ω is the longitude of the perihelion measured from the node, a is the major semi-axis which defines of the orbit of revolution, e is the eccentricity which describes the shape of the orbit, T is the time of perihelion passage defining the position of the body at any time.

Since the functions Ψ and Φ are highly transcendental the solution of equation (1.1) is not obtained by ordinary processes and is very complex as mentioned by Newton. The fundamental elements described by Newton were explained by Laplace [2], in 1780, that also described the method of solution of equation (1.1) that has been the basis for the later works.

In 1809 Gauss [3], in *Theoria Motus Corporum Celestium*, elaborated a technique, based on the least squares, that led to the recovery of the orbit of Ceres lost to sight when its light vanished in the ray of the Sun. Before them Euler [4] in 1744, in *Theoria Motum Planetorum et Cometarum* had obtained an approximate solution while in 1778 Lagrange [5] obtained an original solution which has never been used in practice.

Thus from our point of view the first Dynamic Data Assimilation was made by Laplace and Gauss. In Moulton's book [6], the methods they had adopted are fully described to determine the orbit of a comet. Furthermore in the Gauss method it lays the groundwork for the least squares method that became the foundation of the Data Assimilation.

Previously, in 1805, Adrien-Marie Legendre [7] also had developed the method of least squares, and introduced it in his book *Nouvelles méthodes pour la détermination des orbites des comètes* (New Methods for Determining the Orbits of Comets). The least squares method, independently established by Legendre and Gauss in 1800, also represents the passage from the determinism to the probability.

The basic assumption of physics from the time of Newton up to the Gauss period (late 1600 to 1800) was the determinism. It establishes that the future state of a system is entirely determined by the present state of the system. The evolution of the system is governed by causal relationships as those described, for example, by the Newton equations of motion.

Around the late 1600 the important concept of the probability arose, as a measure of the likeliness that an event will occur. The first to demonstrate the efficacy of defining odds as the ratio of favorable to unfavorable outcomes was Gerolamo Cardano [8] in his book, written around 1564, but the mathematics of the probability was set by Jacob Bernoulli [9], one of the twelve many prominent mathematicians in the Bernoulli family, in the book *Ars Conjectandi* (posthumous, 1713) and by Abraham de Moivre's *Doctrine of Chances* (1718).

In 1812, Laplace [10] published his *Théorie analytique des probabilités* in which it was allowed to compare the merits of different parameter values. For the first time the concept of estimation was introduced, in which acceptable values for the parameters of distributions, specified by hypotheses, are sought. The first half of this treatise was concerned with probability methods and problems, the second half with statistical methods and applications. The term *inverse probability* appeared in the 1837 paper of De Morgan [11] since Laplace's book did not use this term. The term *inverse* means that the probabilities of causes, or hypotheses, could be deduced from the frequency of events. It involves inferring backward from the data to the parameter or from effects to causes. If no information is available when setting initial priors, merely defined as uniform prior, one sets all possible hypotheses to an equal initial prior probability. This concept was introduced by Bayes [12] before and after Laplace [2] that stated the principle:

If an event can be produced by a number n of different causes, then the probabilities of these causes, given the event, are to each other as the probabilities of the event given the causes, and the probability of the existence of each of these is equal to the probability of the event given that cause, divided by the sum of all of the probabilities of the event given each of these causes.

Gauss combined the Bernoulli idea about the probability with Laplace's principle of inverse probability maximized the posterior density of the location parameter in the error distribution, assuming that the prior distribution is uniform. Requiring that the posterior mode is equal to the arithmetic mean, Gauss derived the normal distribution and thus gave a probabilistic justification for the method of least squares (Hald [13]).

By the mid-1700s, the problems of observational error were mathematically described. They were supposed to be positive and negative, and it was generally accepted that their frequency distribution followed a smooth symmetric curve.

The inverse probability describing the probability distribution of an unobserved variable, is called Bayesian probability, and the problem of determining an unobserved variable (by whatever method) is called inferential statistics. The distribution of an unobserved variable given data is rather defined by the likelihood function (which is not a probability distribution), and the distribution of an unobserved variable, given both data and a prior distribution, is the posterior distribution. The term *likelihood* was introduced by Fisher [14] who wrote:

What has now appeared is that the mathematical concept of probability is ... inadequate to express our mental confidence or diffidence in making ... inferences, and that the mathematical quantity which usually appears to be appropriate for measuring our order of preference among different possible populations does not in fact obey the laws of probability. To distinguish it from the probability, I have used the term "Likelihood" to designate this quantity; since both the words likelihood and probability are loosely used in common speech to cover both kinds of relationship (Laplace, 1774, in Stigler's translation [15]).

The development of the field and terminology from *inverse probability* to *Bayesian probability* is described by Fienberg [16] in his paper, where he traces the passage from "inverse probability" concept, due to Bayesian probability, and the birth of the frequentist methods.

At the time that Ronald Alymer Fisher began his studies of statistics at Cambridge in 1909, inverse probability was an integral part of the subject he learned (c.f. Edwards [17]*). Frequentist and other non-Bayesian ideas were clearly "in the air," but it is difficult to know to what extent Fisher was aware of them.*

and more

Fisher's work had a profound influence on two other young statisticians working in Pearson's laboratory at University College London: Jerzy Neyman, a Polish statistician whose early work focused on experiments and sample surveys, and Egon Pearson, Karl Pearson's son. They found Fisher's ideas on significance tests lacking in mathematical detail and, together, they set out to extend and "complete" what he had done. In the process, they developed the methods of hypothesis testing and confidence intervals that revolutionized both the theory and the application of statistics.

Although Fisher disagreed with them, their approaches supplanted inverse probability and later was referred as *frequentist* methods.

There are some differences between Bayesian and frequentist. The frequentist sees the probability as the long-run expected frequency of occurrence while for the Bayesian the probability is related to the degree of belief. It is a measure of the plausibility of an event given incomplete knowledge. For a frequentist the population mean is real, but unknown, and unknowable, and can only be estimated from the data. Knowing the distribution for the sample mean, the frequentist constructs a confidence interval, centered at the sample mean. Since the true mean being a single fixed value, does not have a distribution, and the frequentist cannot say if there is a 95 % probability that the true mean is in a particular interval. The sample mean does. Thus, the frequentist must use circumlocutions like 95 % of similar intervals would contain the true mean, if each interval were constructed from a different random sample like this one. The Bayesian has an altogether different view because for them only the data are real. The population mean is an abstraction, and thus some values are more believable than others, based on the data and their prior beliefs, even though sometimes the prior belief is very non-informative. The Bayesian constructs a credible interval, centered near the sample mean, but adjusted by their *prior* beliefs concerning the mean.

According to David [18] the term "Bayesian," was introduced by R.A. Fisher in 1950 in his *Contributions to mathematical Statistics.*

This short paper to the Cambridge Philosophical Society was intended to introduce the notion of "fiducial probability", and the type of inference which may be expressed in this measure. It opens with a discussion of the difficulties which had arisen from attempts to extend Bayes' theorem to problems in which the essential information on which Bayes' theorem is based is in reality absent, and passes on to relate the new measure to the likelihood function, previously introduced by the author, and to distinguish it from the Bayesian probability a posteriori.

Hald [13] has shown that the method of maximum likelihood was proposed by Daniel Bernoulli, whose paper on likelihood was reported by Kendall [19], but with no practical effect because the maximum likelihood equation for the error distribution was considered intractable. In English the word *likelihood* has been distinguished as being related to, but weaker than, *probability* since its earliest uses. The comparison of hypotheses by evaluating likelihoods has been used for centuries. Different authors by also used the term likelihood, from Christiaan Huygens to Charles Sanders Peirce [20], where model-based inference (usually abduction but sometimes including induction) is distinguished from statistical procedures based on objective randomization.

Fisher [21] also uses the term *method of maximum likelihood estimation (MLE)*. In statistics, maximum-likelihood estimation is a method of estimating the parameters of a statistical model. When applied to a data set and given a statistical model, maximum-likelihood estimation provides estimates for the model's parameters.

The method of maximum likelihood corresponds to many well-known estimation methods in statistics. Assuming that the heights are normally (Gaussian) distributed with some unknown mean and variance, the mean and variance can be estimated with MLE while only knowing the heights of some sample of the overall population. MLE would accomplish this by taking the mean and variance as parameters and finding particular parametric values that make the observed results the most probable (given the model).

With the advent of the statistical thermodynamics due to James Maxwell (see Myrvold [22] and Ludwigh Boltzman [23]) appeared to clear the limitations of the deterministic view. At the same time, however, Poincarè [24], Birkhoff (see Marston Morse [25]) and Lyapunov [26] continued their exploration along the lines of what was called dynamic system whose most recent approach was given by Edward Lorenz [27] with the predictability: the butterfly effect and the strange attractors.

Markov [28], Wiener (see Masani [29]) and Kolmogorov [30] developed the stochastic dynamic prediction related to process that exhibited randomness, in addition to the dynamical system. In probability theory, a stochastic process, or sometimes random process (widely used) is a collection of random variables, representing the evolution of some system of random values over time. This is the probabilistic counterpart to a deterministic process (or deterministic system). Instead of describing a process which can only evolve in one way (as in the case, for example, of solutions of an ordinary differential equation), in a stochastic or random process there is some indeterminacy: even if the initial condition (or starting point) is known, there are several (often infinitely many) directions in which the process may evolve.

In the simple case of discrete time, as opposed to continuous time, a stochastic process involves a sequence of random variables and the time series associated with these random variables (for example, see Markov chain, also known as discrete-time Markov's chain). One approach to stochastic processes treats them as functions of one or several deterministic arguments (inputs, in most cases regarded as time) whose values (outputs) are random variables: non-deterministic (single) quantities which have certain probability distributions. Random variables corresponding to various times (or points, in the case of random fields) may be entirely different. The main requirement is that these different random quantities all have the same type. Type refers to the codomain of the function. Although the random values of a stochastic process at different times may be independent random variables, in most commonly considered situations they exhibit complicated statistical correlations.

Familiar examples of processes modeled as stochastic time series include stock market and exchange rate fluctuations, signals such as speech, audio and video, medical data such as a patient's EKG, EEG, blood pressure or temperature, and random movement such as Brownian motion or random walks. Examples of random fields include static images, random terrain (landscapes), wind waves or composition variations of a heterogeneous material. By mid-nineteenth, there was a renewed interest on Bayesian approach that laid the groundwork for filtering theory, smoothing, prediction by Kolmogorov [30], Kalman and Bucy [31].

At the end of our more recent history, we have the work on predictability, data analysis via minimum variance and in the frame of quantum studies the Monte Carlo approach by Metropolis and Ulam [32].

This area of research is one of the most active and one of the most challenging in the frame of ensemble prediction. There are a variety of avenues that are being explored and that will be shown in this book. Among them one can cite the ensemble Kalman filter (see Evensen and van Leeuwen [33], Burgers et al. [34], Hamill et al. [35], Evensen [36]).

Lewis [37] offers a complete picture of the genealogy of data assimilation constructed using the information coming from the Dictionary of Scientific Biography (Gillespie [38]). Such a chart has been drawn here with some small differences (see Fig. 1.1).

1.2 Approaching the Meteorological System

With the invention of the first meteorological instruments the quantitative observation of meteorological phenomena in 14th century was initiated. The first regular instrumental observation was planned by Blaise Pascal at Puy de Dôme in Paris where he observed the first relation between the weather change and the height of mercury column in the thermometer. The first systematic observations were also made in London and Oxford. Previously non-systematic observations had been also made in Florence by Galileo and his scholars, including Torricelli [39], in the first climate network set up by Duke Ferdinand II of Tuscany (1610–1670). The first observations in

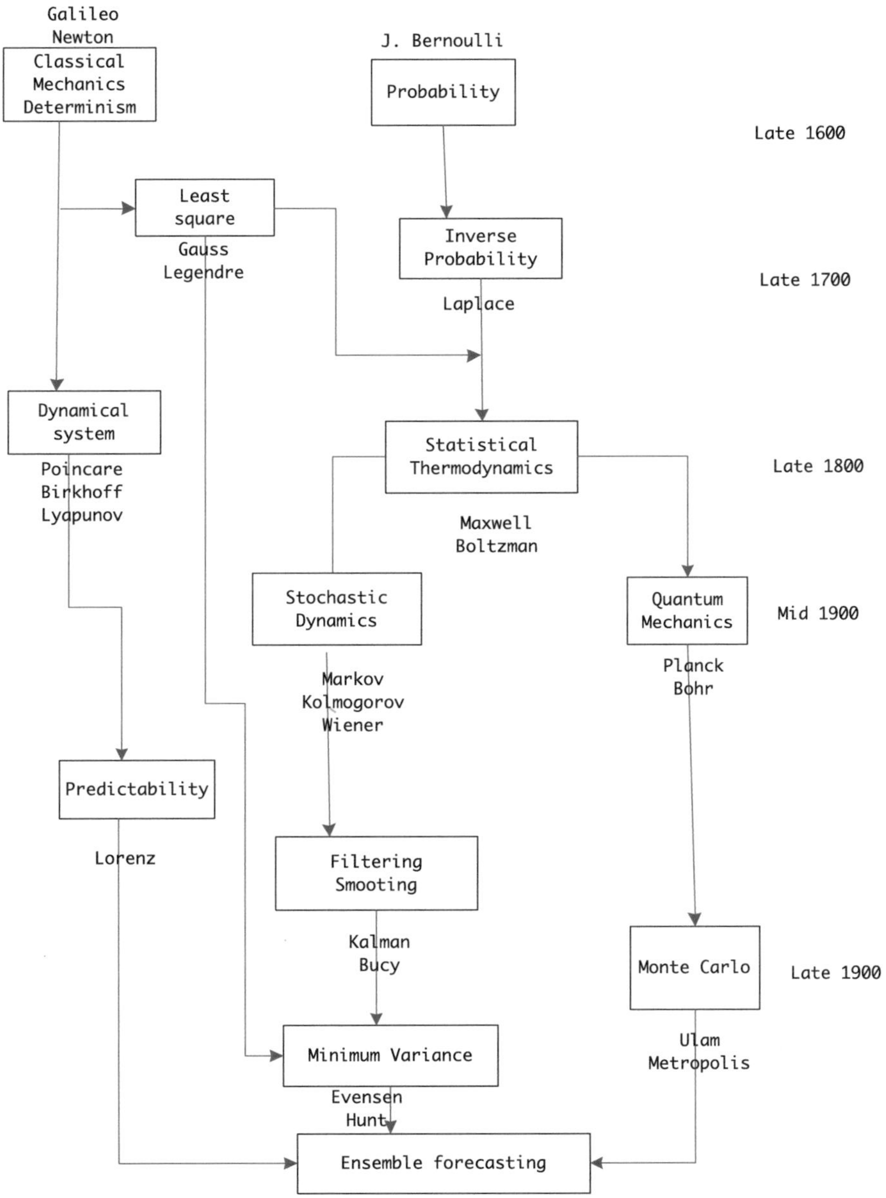

Fig. 1.1 I have used part of this figure to draw the structure of this book, changing it in the early part when probability studies arose. *Source* Lewis [37]

Germany started in 1654 when the Jesuit college of Osnabröak received a Florentine thermometer. During the first part of the 18th century systematic observations were begun in Russia. Due to the heterogeneity of the instruments and mainly because they were not standardized and calibrated it was impossible to make some forecasts.

In 1763 in Mannheim *The Palatine Academy of Sciences and Letters'* was founded that in 1780 established a special meteorological class of the Academy to be known as the Palatine Meteorological Society (Societas Meteorologica Palatina). Many observatories participated in the observation proposed by the Mannheim Society. The observations collected were published in Latin, which was the international language of the science, in the journal of the Society: *Ephemerides*. Nevertheless the Mannheim Society had closed its activities in 1799, the concept had been initiated and the first attempt to standardize the observations could be extended to all observations on ground stations everywhere.

The present meteorological network has evolved over 300 years with new instruments and new technologies. With the advent of satellites, the instruments of class 1, which are the instruments making measurements at points, were joined by class 2 instruments, that are instruments sampling an area or a volume remotely. A successive class 3 devices were also introduced that are instruments calculating the wind velocities from Lagrange trajectories.

While the instrumental error for traditional class 1 instruments can be defined within a certain accuracy, the instrumental errors for the class 2 and 3 instruments are not so straightforward because their results are derived from inversion methods and therefore are strongly dependent on the algorithms used.

Another type of error was also reported: the error of representativeness. Petersen and Middleton [40] defined a representative observation as a datum pertaining to a particular point and time, which is the result of an optimum filtering operation on the continuous raw data field, under the criterion of minimum average mean square error of reconstruction from the subsequently sampled values on a given space-time lattice.

Starting from this definition one can say that the error of representativeness is a measure of the error due to misrepresentation in space-time scale of the observation. Such error can arise from the non-correct distance of the stations in the observation network, in case of hazard or natural phenomena, or in case of micro observations, in a non-sufficient density of observations able to capture the essence of phenomenon under investigation.

Petersen [41] defines the expression of the observation process as a set of measurements of the physical field made at various space-time points by means of devices having certain physical and statistical features. The main refinement is that the totality of this information be instantaneous in time and, whether counted in units of real numbers, be finite. The criterion by which these estimates are to be evaluated is their mean square error or deviation from the true value of the field variable at the given location and time. Then the optimal solution to this problem is that the estimate should coincide with the mean value or expectation of the variable, conditional upon the available data.

Surface stations, radiosondes and pilot balloons are irregularly distributed over land and on some remote islands. Then they are complemented with meteorological satellites Meteosat (Europe), GOES (USA), Elektro-L1 (Russia), MTSAT-1R (Japan), INSAT (India), Feng-Yun (China) and Earth Observation Satellite constellations as for example the NASA A-Train or the near future European, ESA, constellation named Sentinel.

These constellations provide data that can be used together to obtain comprehensive information about atmospheric components or processes that are happening at the same time. Combining the information collected simultaneously from several sources, by the ground and airborne network, one obtains a complete answer to many questions than would be possible from any one satellite taken alone at different times.

Data assimilation methods used in Numerical Weather Prediction, where grid measurements are present and model process are available, have also been developed in other fields: earthquakes, chemical process in atmosphere, planetary circulation, robotics and biology growth. The applications refer to methods for updating the state vector (initial condition) of a complex space-time model by combining new observations with one or more prior forecasts, in which the changing of the space-time observations requires proper sampling and the observation representativeness and related errors become crucial to both follow and forecast the phenomenon under investigation.

Reviewing the concepts expressed above it is possible to define to better what is the assimilation and its cycle. It can be defined as

1. data checking;
2. proper interpolation of data;
3. initialization of the assimilation model;
4. short forecast to prepare the background field.

Let us now reverse our approach and look first of all at the assimilation model instead of observation data. In general a dynamic system is described by a system of non-linear partial equation (PDEs) whose solution is discrete. If the solution of the system is well-posed in the sense of Hadamard, there is a forward discrete operator that yields the solution. Let us note that a dynamic system evolves with time but that not all of the state variables, within the state vector, have equal information content and not all state variables are known to the same precision. It is therefore desirable that the observations made both contain the maximum information content possible and allow the systems state to be characterized with a minimum uncertainty. Furthermore, in order to pass from the observation location to the grid points of the model some interpolations are performed. In fact the first-guess fields defined at the grid points of the forecasting model are interpolated to the observation location, while differences between the observation and the interpolated value are then interpolated back onto the grid points to define a correction.

Although the continuum state of the observations is the most natural candidate to be accumulated into a model, the true state to be estimated is a projection of the continuum state on a discrete space. This refers to the best possible state represented by the model, which is what we are trying to approximate. Thus the production of an accurate image of the true state of the system at a given time, represented in a model as a collection of numbers, is called analysis. An analysis can be useful in itself as a comprehensive and self-consistent diagnostic of the system. It can also be used as input data to another operation, notably as the initial state for a numerical forecast, or as a data retrieval to be used as a pseudo-observation. It can provide a reference against which to check the quality of observations.

There are two basic approaches: sequential assimilation, which only considers observation made in the past until the time of analysis, which is the case of real-time assimilation systems, and non-sequential, or retrospective assimilation, where observation from the future can be used, for instance in a reanalysis exercise.

Since the model has a lower resolution than reality, even the best possible analysis will never be entirely realistic. Thus even though the observations do not have any instrumental error, and the analysis is equal to the true state, there will be some unavoidable discrepancies between the observed values and their equivalents in the analysis, because of *representativeness errors*. Although we will often treat these errors as a part of the observation errors in the mathematical equations, one should keep in mind that they depend on the model discretization, not only on instrumental problems.

Summarizing, the necessary objective information that we can use to produce an analysis is a collection of observed values provided by observations of the true state. If the model state is overdetermined by the observations, then the analysis reduces to an interpolation problem. In several cases, the analysis problem is under-determined because data are sparse and only indirectly related to the model variables, as happens in remote sensing measurements. In order to make it a well-posed problem it is necessary to rely on some background information in the form of an a priori estimate of the model state. Physical constraints on the analysis problem can also help. The background information can be a trivial state; it can also be generated from the output of a previous analysis, using some assumptions of consistency in time of the model state, like stationarity (hypothesis of persistence) or the evolution predicted by a forecast model.

1.3 Numerical Weather Prediction Models

An interesting paper by Lewis [37] outlines the history of the numerical weather forecasting, from the first deterministic solutions up to the most recent approaches based on Monte Carlo methods.

Around 1950, it was clear that the dynamics of weather could be evaluated using the same equation of dynamics developed in other fields. The theoretical treatment of the scales of motion in the atmosphere, initially based on the determinist set of equations, proposed by Jule Charney [42, 43] showed be considered the first successful Numerical Weather Prediction (NWP). At Princeton's Institute for Advanced Study under the guidance of John von Neumann, Jule Charney [43] and his team made two successful 24 h forecasts of the transient features of the large-scale flow, initialized on 30 January and 13 February 1949, even though the successive 24 h forecasts made (5 January 1949, e.g.) were not particularly useful, as referred by Lewis [37] in his paper.

The reactions of the meteorologic community were positive even though it was clear the limits of deterministic prediction that were governed by the growth of errors because the solution depends both on the initial state, which is generally erroneous, and on the models that are, by their nature, imperfect. Furthermore the feature of the causal laws, characterized by unstable systems and non-periodicity limited the predictability of the system.

Whereas Gauss and contemporaries had found that the two-body problem of celestial mechanics tolerated, in the initial state, small errors, meteorological prediction under non-periodic constraints would be found to be less forgiving of these uncertainties [37].

In the first years, the major theme was the development of the models and, of course, the identification and related correction of systematic errors. That the deterministic forecast was imperfect had been already discovered by Eady [44] in his *Compendium of meteorology*.

...we never know what small perturbations may exist below a certain margin of error. Since the perturbations may grow at an exponential rate, the margin of the error in the forecast (final) state will grow exponentially as the period of the forecast is increased, and this possible error is unavoidable whatever our method of forecasting ... if we are to glean any information at all about developments beyond the limited time interval, we must extend our analysis and consider the properties of the set or ensemble (corresponding to the Gibbs-ensemble of statistical mechanics) of all possible developments. Thus, long range forecasting is necessarily a branch of statistical physics in its widest sense: both our questions and answers must be expressed in terms of probabilities.

In 1962 Edward Lorenz [45], using a truncated version of the two levels quasi-geostrophic model described by Lorenz himself [46], found that:

However the work the existence of deterministic systems governed by equations whose nonlinearity resembles the non-linearity of the atmosphere, which are not perfectly nor almost perfectly predictable by simple and simply determined linear formulas, if the period between successive observations is greater than half of the shortest significant period of observation.

Lorenz speculated that, by a linear regression, one day forecast should be good, but successive days forecast should be poor. An unexpected result occurred when Lorenz inadvertently introduced truncation errors in the model he was running. The small error had influenced the three decimal place, instead of six decimal but had amplified so much of the simulation that the signal had been covered *...and I found this very exciting because this implied if the atmosphere behaved this way, the long range forecasting was impossible ...*

The contribution by Lorenz [27], using a system of three ordinary differential equations of a simplified system of Barry Saltzman to study finite amplitude convention, has laid the foundation for the field of chaotic systems (see Chap. 5).

Despite Lorenz's numerical experiments not being encouraging, scientists continued to make weather experiments for intermediate time, that was of the order of weeks.

The Global Atmospheric Research Program of 1968 [47] became the reference Program, and the General Circulation Model (GCM) became the instrument for the experiments.

In the meantime, an alternative to the classical Numerical Weather Prediction models was also developed. It was based on a stochastic dynamic approach that coupled the probability with the determinism whose practical implementation in meteorology was called ensemble prediction.

We can follow the reaction of Lorenz twho in 1965 claimed: *Among the possible internal mechanism for the production of long-term fluctuations, one of the most interesting to consider is due to the non-linearity of the governing physical law. It seems possible that there may exist a number of distinct régimes, such that the general circulation, upon finding itself in any particular régimes, tends to become stuck there. That is, the atmosphere may readily progress from one state to another within a given régime, but only occasionally acquire just such a state as to allow it to pass to one rŕegimes to another. The change of régimes will then appear as long-period fluctuates.*

and moreover:

The aperiodicity of the variations indicates that if the initial conditions are not precisely known, prediction at a sufficiently long range is impossible, no matter how well the governing law may be formulated. If distinct rŕegimes are present it may be possible to predict the rŕegimes with a reasonable probability of success, at a considerably longer range than that at which one can hope to predict the state within the rŕegime. The proposed procedure chooses a finite ensemble of initial states, rather than the single observed initial state. Each state within the ensemble resembles the observed state closely enough so that the differences might be ascribed to errors or inadequacies in observation. A system of dynamic equations previously deemed to be suitable for forecasting is then applied to each member of the ensemble, leading to an ensemble of states at any future time. From an ensemble of future states, the probability of occurrence of any event, or such statistics as the ensemble mean and ensemble standard deviation of any quantity, may be evaluated. Between the near future, when all states within an ensemble will look about alike, and the very distant future, when two states within an ensemble will show no more resemblance than two atmospheric states chosen at random, it is hoped that there will be an extended range when most of the states in an ensemble,while not constituting good pin-point forecasts,will possess certain important features in common. It is for this extended range that the procedure may prove useful.

In the early 1960s, Edward Epstein, a professor at University of Michigan, began to use probabilistic methods called Monte Carlo [32] defined by Metropolis and Ulam. While attending an American Meteorological Society meeting of 1968, in listening to Lorenz's presentation he was influenced in such a way as to redirect his work. It was clear that the initial position of the motion was subject to indeterminacy, but one started, the motion followed the deterministic law. *The variables were not considered deterministic variables, rather as random variables with associated probabilistic, that is, stochastic properties* [37]. During a sabbatical year at the International Institute of Meteorology in Stockholm in 1968, he began work on a stochastic dynamic (SD) approach to Numerical Weather Prediction which culminated with the paper: *Stochastic dynamic prediction* [48].

1.4 What, Where, When

Assimilation is a word that covers different meanings, from those linguistic to those cultural, from the process of conversion of nutrient in biology to those incorporating new concepts into existing schemes as it is in psychology, up to the technical assimilation in meteorology or climate, robotics and recently biology processes. In such cases regarding the word assimilation, we need to append the word data. One shall cover this last aspect analyzing the dynamic data assimilation.

In this frame the Data Assimilation is a set of mathematical techniques allowing us to use all the information available to us within a time frame, including observational data, any prior information we may have by a deterministic model describing our system and encapsulating our theoretical understanding of Data Assimilation. The mathematical basis is the estimation theory or the theory of inverse problems that is an organized set of mathematical techniques for obtaining useful information about the physical world on the basis of observations.

In a conventional problem, one would use a set of known prior parameters to predict the state of a physical system. This approach is usually called a forward problem, whereas in the inverse problem one attempts to use available observation of the state of the system to estimate poorly known parameters of the state itself. In both the case, Data Assimilation can be treated as a Bayesian system.

The Bayesian theorem or the law of inverse probability allows us to combine prior information about the parameters with the information contained in the observations, to guide the statistical inference process. The reason why the Data Assimilation is so effective is that it seeks to produce an analysis that fits a set of observations taken over a time frame, not just the observations made at one instant in time, subject to the strong constraint that the evolution of the analyzed quantities is governed by a deterministic model describing the given observation.

Data Assimilation has been applied to several fields, from the numerical weather prediction, where it began, to planetary climate analysis up to the evolution of biological cells or in a robotic system. One begins with a forecast model, often called background. In order to make a useful prediction, the background must frequently be updated with noisy and sparse measurements. This procedure updates the background in light of the new observation to produce an analysis, which, under suitable assumptions is the maximum estimate of the model state vector. Subsequently, the model is restarted from analysis and provides a new background or forecast. Data Assimilation and model forecast can be combined into an observing system simulation experiment quantifying the effect of changes in the observation accuracy, type, location and frequency on the accuracy of the numerical forecast.

In the case of weather forecasting, like many other systems, one cannot sample the atmosphere at every point, observations are corrupted by noise, and any given model is imperfect.

For these reasons the method suffers from some limitations, because eventually the forecasted parameters diverge from the true values which have subsequently introduced, the optimal procedures combine the most exhaustive theoretical and observational knowledge with the errors of observation and those due to the model. Furthermore dynamic systems exhibit sensitive dependence on initial conditions. Those are small errors in non-fixed-point initial conditions quickly propagated in time, leading to substantial differences in solutions. Even though trajectories have similar limit sets, they became uncorrelated over time even when the initial conditions were very similar and when, on time scales of a few days or less, uncertainties in the initial state of the atmosphere may lead to substantial forecast errors.

In a classic example, consisting of a system of three coupled ordinary differential equations modeling fluid flow, Edward Lorenz [27] found a chaotic behavior that led to models that did not offer any predictive advantage over climatological averages. To account for uncertainties in a forecast model, Lorenz suggested that instead of simulating a single initial condition, under the model from a best estimate of the state of the atmosphere, one should evolve a set or ensemble of initial conditions, each from a statistically equivalent estimate of the true initial state, that assesses and quantifies the effect of uncertainty in a mathematical model of a dynamical system. The ensemble gives a Monte Carlo estimate of the uncertainty in a given dynamical model. Under assumptions to be discussed in the next sections, the ensemble mean constitutes an empirical maximum-likelihood estimate of the true state of the atmosphere.

Ensemble forecasting, the technique used in the frame of the early studies of chaotic behavior is now part of the routine operations at the U.S. and European weather centers.

Summarizing Data Assimilation is an analysis in which the information is stored in a dynamic model, exploiting the consistency constraints inherent in physical laws and processes, combining temporal observations distributed over time with the dynamic model itself. The analysis process is to varying degrees:

1. an approximation of the true state of a physical system at a given time;
2. a comprehensive diagnosis and consisting of a physical system;
3. a reference by which to check the quality of the observations;
4. a matter of useful inputs for another task, such as the initial state of a predictive model.

A paradigmatic case is one of the assimilation of a thunderstorm forecast, which implies that errors due to initial conditions must be reduced as far as possible leaving only the model to generate errors and then proceed in a realistic direction. In this way, it will develope not only the prediction, but also searching for a collection of accurate data and diagnosis of errors of the model itself. The acquisition combines the observation data with the data produced by the model to reproduce an *optimal* estimate evolving state of the system. The model provides texture to observed data allowing interpolating or extracting them in regions of space and time in which they are missing. Moreover, the observed data adjust the trajectory of the model through the state of the model itself, keeping it *online* in a loop of prediction-comment-correction.

There are at least two basic approaches for data assimilation: one where we consider sequential data observed in the past until the time of writing, in this case, we have similar systems in real time, and one non-sequentially where one can use successive observations. Methods may be intermittent or continuous over time.

Temporal intermittent distribution includes a cycle of 6 h and observations are processed in batches. Continuous observations are carried out on a long run and analyzed. The state correction is smoothed over time allowing one to obtain a more realistic analysis.

From a practical point of view, one uses a template, the so-called direct model, to connect the input parameters to the output parameters. Mathematically one writes:

$$\mathbf{y} = \mathbf{H}(\mathbf{x}) \tag{1.2}$$

where $\mathbf{x}$ represents the collection of all the variables that describe the state of the model. In this way one can compare observations obtained from the model with those obtained by $\mathbf{y}$, estimating the error of the model itself. The measured data can come from different sources, from in situ measurements or from satellite. Then the numeric features of assimilation are reduced to a minimization problem where the cost function $\mathcal{J}$ is:

$$\mathcal{J} = ||\mathbf{y} - \mathbf{H}(\mathbf{x})||^2 \tag{1.3}$$

where $|| \cdot ||$ is the norm two.

Data Assimilation is applied in several fields. From the first applications, as those developed as an example, by Gauss around the reappearance of the planetoid Ceres, as it is reported in previous paragraphs, the techniques were mainly developed in Numerical Weather Prediction and became mature with a right blended of deterministic and stochastic approach.

Recent developments have been made in covering new fields like the planetary atmospheric circulation, mainly Martian atmosphere and in a field apparently far from the current data assimilation like that devoted to bio-medicine where one addresses the problem of how differences between the predicted state of a biological system can be reconciled with noisy measurements to correct the forecast in view of new information.

References

1. Newton I.: Principia book III prop XLI. https://archive.org/details/newtonspmathema00newtrich
2. Laplace, P.S.: Mémoire sur la probabilité des causes par les évènements," Mémoires de l'Académie Royale des Sciences Presentés par Divers Savans **6**, 621–656 (1774). Translated in S. Stigler Laplace's 1774 Memoir on Inverse Probability, Statistical Science **1**, 359–378 (1986)
3. Gauss, C.F.: Theoria motus corporum celestium, Hamburg: Perthes und Besser (1809). Translation by C.H. Davis in Theory of Motion of Heavenly Bodies. Dover, New York (1963)
4. Euler, L. : Theoria Motum Planetorum et Cometarum https://math.dartmouth.edu/~euler/tour/tour_17.html
5. Lagrange, J.L. (1776). Mèmoire sur l'utilité de la méthode de prendre le milieu entre les résultats de plusieurs observations; dans lequel on examine les avantages de cette méthode par le calcul d es probabilités, ou l'on resoud differens problémes relatifs à cette matiére. Miscellanea Taurinensia 5 167–232. Reprinted in Lagrange. Oeuvres de Lagrange, 2. Gauthier-Villars, Paris. (1868)
6. Moulton, F.R.: An Introduction To Celestial Mechanics. Kessinger Pub Co (2007)
7. Legendre, A.M.: Sur la Méthode des moindres quarrés. On the Appendix of Nouvelles méthodes pour la détermination des orbites des comètes. Courcier, Paris (1805)
8. Cardano G.: Liber de ludo aleae (Book on Games of Chance). Princeton University press (1953)
9. Bernoulli, J.: Ars Conjectandi. Thurnisiorum, Basel (1713)
10. Laplace, P.S.: Théorie analytique des probabilités, 1st edn., 1812, 3rd ed., with supplements. Courcier, Paris (1820)
11. de Morgan, A.: An Essay on Probabilities and Their Application to Life Contingencies and Insurance Offices. Longmans, London (1838)
12. Bayes, T.: An essay towards solving a problem in the doctrine of chances. Philos. Trans. R. Soc. Lond. for 1763 **53**, 370-414 (1764). Reprinted with an introduction by G.A. Barnard In: Pearson, E.S., Kendall, M.G. (eds.) Studies in the History of Statistics and Probability, pp. 131-153. Chas. Griffin, London (1970)
13. Hald, A.: A History of Mathematical Statistics from 1750 to 1930. Wiley, New York (1998)
14. Fisher, R.A.: Statistical Methods, Experimental Design, and Scientific Inference, Being a Reprint of Statistical Methods for Research Workers (1925). The Design of Experiments (1935), and Statistical Methods and Scientific Inference (1956). University Press, Oxford (1990)
15. Stigler, S.M.: Laplace, Fisher, and the discovery of the concept of sufficiency. Biometrica **60** 439–445. Reprinted in 1977. In: Kendall, M.G., Plackett, R.L. (eds.) Studies in the History of Statistics and Probability, vol. 2, pp. 271–277. Griffin, London (1973)
16. Fienberg, S.E.: When Did Bayesian Inference Become "Bayesian"? Bayesian Anal. **1**(1), 1–40 (2006)
17. Edwards, A.W.F.: What did Fisher mean by "inverse probability" in 1912–1922. Stat. Sci. **12**, 177–184 (1997)
18. David, H.A.: First (?) occurrence of common terms in mathematical statistics. Am. Stat. **49**, 121–133 (1995)
19. Kendall, M.G.: Studies in the history of probability and statistics XI Daniel Bernoulli on maximum likelihood. Biometrica **48**(1–2), 1 (1961)
20. Pierce A. : Theory of Probable Inference in Studies in Logic by Members of the Johns Hopkins University (1883)
21. Fisher, R.A.: Contributions to Mathematical Statistics. Wiley, New York (1950)
22. Myrvold W.C.: Statistical mechanics and thermodynamics: a Maxwellian view. Stud. Hist. Philos. Mod. Phys. **42**, 237–243 (2011)
23. Boltzmann, L.: Vorlesungen über Gastheorie: 2 Volumes—Leipzig 1895, 98 UB: O 5262–6 (1896–1898). English version: Lectures on gas theory. Translated by Stephen G.B.: University of California Press, New York (1964). ISBN: 0-486-68455-5 (1995)

24. Poincaré, H.: The Foundations of Science. Science Press, New York. Reprinted in 1921 (1902-1908); This book includes the English translations of Science and Hypothesis (1902), The Value of Science (1905), Science and Method (1908)
25. Marston, M.: Bull. Amer. Math. Soc. **52**(5), Part 1, 357–391 (1946) (see project euclid)
26. Lyapunov, A.M.: The general problem of the stability of motion. Translated by Fuller, A.T. Taylor and Francis, London. ISBN: 978-0-7484-0062-1. Reviewed in detail by Smith, M.C.: Automatica 1995 **3**(2), 353–356 (1992)
27. Lorenz, E.N.: Deterministic non-periodic flow. J. Atmos. Sci. **20**, 130–141 (1963)
28. Markov, A.A.: Extension of the limit theorems of probability theory to a sum of variables connected in a chain. Reprinted in Appendix B of: Howard, R. Dynamic Probabilistic Systems, vol. 1, Markov Chains. Wiley (1971)
29. Masani, P. (ed.): The Mathematical Work of Norbert Wiener, vol. 4. MIT Press, Cambridge. This contains a complete collection of Wiener's mathematical papers with commentaries (1976)
30. Kolmogorov, A.: Grundbegriffe der Wahrscheinlichkeitsrechnung (in German). Julius Springer, Berlin. Translation: Kolmogorov, A. (1956). Foundations of the Theory of Probability, 2nd edn. Chelsea, New York (1933)
31. Kalman, R.E., Bucy, R.S.: New results in linear filtering and prediction theory. Trans. Am. Soc. Mech. Eng., J. Basic Eng. Ser. D **83**, 95–108 (1961)
32. Metropolis, N., Ulam, S.: The Monte Carlo method. J. Am. Stat. Assoc. **44**(247), 335–341 (1949)
33. Evensen, G., van Leeuwen, P.: Assimilation of Geosat altimeter data for the Agulhas current using the ensemble Kalman filter with a quasi geostrophic model. Mon. Wea. Rev. **124**, 85–96 (1996)
34. Burgers, G., van Leeuwen, P.J., Evensen, G.: Analysis scheme in the ensemble Kalman filter. Mon. Wea. Rev. **126**, 1719–1724 (1998)
35. Hamill, T., Mullen, S., Snyder, C., Toth, Z., Baumhefner, D.: Ensemble forecasting in the short to medium range: report from a workshop. Bull. Am. Meteor. Soc. **81**, 2653–2664 (2000)
36. Evensen, G.: Data Assimilation: The Ensemble Kalman Filter. 2nd edn. p. 320. Springer (2009)
37. Lewis, J.M.: Roots of ensemble forecasting. Am. Meteorol. Soc. **133**(7), 1865–1885 (2005)
38. Gillispie, C. (ed.): Dictionary of Scientific Biography, vol. 18. Scribner, New York (1981)
39. Torricelli, E.: http://www-gap.dcs.st-and.ac.uk/~history/Mathematicians/Torricelli.html
40. Petersen, D.P., Middleton, D.: On representative observations. Tellus XV **4** (1963)
41. Petersen, D.P.: On the concept and implementation of sequential analysis for linear random fields. Tellus XX **4**, 673–686 (1968)
42. Charney, J.: On the scale of the atmospheric motions. Geofys. Publ. **17**(2), 17 (1948)
43. Charney, J.: The use of the primitive equations of motion in numerical weather prediction. Tellus **7**, 22–26 (1955)
44. Eady, E.: The quantitative theory of cyclone development compendium of meteorology. In: Malone, T. (ed.) American Meteorological Society, pp. 464–469 (1951)
45. Lorenz, E.N.: The statistical prediction of solutions of dynamic equations. In: Proceedings of the International Symposium on Numerical Weather Prediction, pp. 629–634. Meteorological Society of Japan, Tokyo (1962)
46. Lorenz, E.N.: Energy and numerical weather prediction, Tellus XII **4**, 364–373 (1960)
47. GARP: GARP topics. Bull. Am. Meteor. Soc. **50**, 136–141 (1969)
48. Epstein, E.: Stochastic dynamic prediction. Tellus **21**, 739–759 (1969)

Chapter 2
Representation of the Physical System

Abstract The core problems of estimation approaches to data assimilation problems lie in the generic discrete stochastic dynamic model of the system components and the generic discrete stochastic model of the observations. Thus, the first step in the mathematical formulation of the analytical problem is the definition of the work space, data quality and the model used to represent the system dynamics. This chapter deals with the estimation problem, the representativeness of the model and the Optimal Estimation techniques. A special attention is also devoted to explaining why the data assimilation is an inverse problem.

2.1 The Observational System and Errors

The core problems of estimation approaches to data assimilation problems lie in the generic discrete stochastic dynamic model of the system components and in the generic discrete stochastic model of the observations. Discrete dynamics are assumed as given and the difference between the discrete dynamics and the governing continuum dynamics is accounted for by model error represented by stochastic forcing. Since the system state is considered discretely whereas it is the continuum state that should be observed, the observation model includes a representativeness error term as well as a measurement error term.

Although the applicable equations imply that a continuous "initial condition" field description is sufficient to define the values of the field for all future time, the uncertainty of the synoptic reconstruction, caused by finite sampling density and inaccurate measurements, suggests that continuity stems from a previous analysis may assist in defining the present distribution of values. The analyst utilizes this principle intuitively by inspecting the current array of data for patterns previously identified [1].

As a natural consequence the design of observation and data collection system for analysis and prediction of physical fields needs to be oriented toward periodic simultaneous measurements throughout the medium under investigation. The analysis consists of a reconstruction of the continuous spatial field, or a dense grid

R. Guzzi, *Data Assimilation: Mathematical Concepts and Instructive Examples*, SpringerBriefs in Earth Sciences, DOI 10.1007/978-3-319-22410-7_2

representation of the same, to which the dynamic equations of motion are applied in order to extrapolate into future time.

Thus the first step in the mathematical formulation of the analytical problem is the definition of the work space. The goal is to find a true state vector that is the projection of the infinite dimensional space of the field vector to the finite dimensional space of its numerical representation.

In dynamic meteorology and related disciplines, the forecasts of physical field variables and the mathematical models of field dynamics are expressed by a system of nonlinear partial differential equations PDEs whose prognostic state variables at time t_k is the vector $\mathbf{x}_k$.

Assuming the governing PDEs to be well-posed in the sense of Hadamard there is a unique solution operator or time dependent propagator $\mathbf{g}$ that yields the solution $\mathbf{x}_k$ given the solution $\mathbf{x}_{k-1}$ at an earlier time interval $t_k - t_{k-1}$. Omitting the time index t_k, from now on represented by the index k, we have:

$$\mathbf{x}_k = \mathbf{g}(\mathbf{x}_{k-1}) \tag{2.1}$$

for $k = 1, 2, 3 \ldots$. It is a faithful representation of the dynamics of the system under investigation.

This system could be stochastically forced, for example through uncertain boundary conditions or also be internally forced by stochastic free parameters estimated during the course of data assimilation, for example by physical parameterization. If the presentation parameters and forcing are considered fixed the propagator $\mathbf{g}$ is deterministic.

Since the numerical representation is defined through a discretization process, one has a discretized version $\mathbf{x}_k^d$ of Eq. 2.1 by the discrete propagator $\mathbf{f}$. How these vector components relate to the real state depend on the choice of discretization, which is mathematically equivalent to a choice of basis.

$$\mathbf{x}_k^d = \mathbf{f}(\mathbf{x}_{k-1}^d), \tag{2.2}$$

for $k = 1, 2, 3 \ldots$, where the superscript d means discrete.

Since the reality is more complex than what can be represented by a state vector, one must distinguish between reality itself and the best possible representation of reality, which one denotes as the *true* state at the time of the analysis. Thus we define a discrete *true* state vector $\mathbf{x}^t$ of dimension n, based on the available observation $t_1, t_2, \ldots$, that is the representation of the discrete state on a continuum state.

$$\mathbf{x}_k^t \equiv \mathbf{\Pi}\mathbf{x}_k \tag{2.3}$$

where $\mathbf{\Pi}$ is an operator mapping, in a proper manner, the discrete space on continuous space. The *true* state is still unknown since $\mathbf{x}_k$ is unknown and also the initial condition $\mathbf{x}_0$ and the propagator $\mathbf{g}$ are unknown.

Using the previously defined operators and applying the operator $\mathbf{\Pi}$ on both side of Eq. 2.1 and adding and subtracting $\mathbf{f}(\mathbf{x}_{k-1}^t)$ we obtain the discrete evolution of the equation for $\mathbf{x}_k^t$

$$\mathbf{x}_k^t = \mathbf{f}(\mathbf{x}_{k-1}^t) + \mathbf{e}_{k-1}^t \tag{2.4}$$

where $\mathbf{f}$ is the discrete propagator that resides in our numerical solution and the forcing term $\mathbf{e}_{k-1}^t$ is the model error from time t_{k-1} to t_k that is:

$$\mathbf{e}_k^t = \mathbf{\Pi g}(\mathbf{x}_k) - \mathbf{f}(\mathbf{\Pi x}_k) \tag{2.5}$$

that is appropriate to be represented as a stochastic perturbation because:

- is state-dependent;
- its dependence on unknown $\mathbf{\Pi}$ renders it, from a deterministic point of view, unknowable;
- is small provided $\mathbf{f}$ approximates $\mathbf{g}$.

Let us now explore the observed data on the basis of the *true* state $\mathbf{x}_k^t$. Le us define a continuous observing vector obtained by a number of time dependent observations, with their errors, made at time t_k where $k = 1, 2, 3, \ldots$. If the error is additive we can write:

$$\mathbf{x}_k^{obs} = \mathbf{h}_k^c(\mathbf{x}_k) + \mathbf{e}_k^m \tag{2.6}$$

Here $\mathbf{h}_k^c$ is the continuum forward observation operator and $\mathbf{e}_k^m$ is the measurement error, considered stochastic, whose mean is: $\hat{\mathbf{e}}_k^m \equiv E[\mathbf{e}_k^m]$. $E[\cdot]$ is the expectation operator.[1]

The observation operator can be considered linear when the state variables are directly observed, as in case devices located in ground measurement stations and radiosondes and non linear when data come from remotely sensed devices that require proper integro-differential algorithms to be interpreted.

[1]The expectation of a random variable is defined as the sum of all values the random variable may take, each weighted by the probability with which the value is taken. In term of formula the expectation of $E[x]$ is given by:

$$E[x] = \int_{-\infty}^{+\infty} x f(x) dx \tag{2.7}$$

This is also called mean value of x or first moment. A second moment is given by the quantity $E[x^2] = \int_{-infty}^{+infty} x^2 f(x) dx$. The variance of a random variable is the mean squared deviation of the random variable from its mean; it is $\sigma^2 = E[x^2] - E[x]^2$. Another important concept is the statistical correlation between random variables that is given by the covariance, which is the expectation of the product of the deviations of two random variables from their means:

$$E[(x - E[x])(y - E[y])] = E[xy] - E[x]E[y] \tag{2.8}$$

that is a measure of bias, and its covariance matrix $\mathbf{R}_k \equiv E[(\mathbf{e}_k^m - \hat{\mathbf{e}}_k^m)(\mathbf{e}_k^m - \hat{\mathbf{e}}_k^m)^T]$.

Let us now formulate the stochastic dynamic model adding and subtracting $\mathbf{h_k}(\mathbf{x}_k^t)$ from the relation (2.6) taking into account the discrete *true* state $\mathbf{x}_k^t = \mathbf{\Pi x}_k$. Then the discrete observation model is obtained by:

$$\mathbf{x}_k^{obs} = \mathbf{h_k}(\mathbf{x}_k^t) + \mathbf{e}_k^{obs} \tag{2.9}$$

where $\mathbf{h}_k$ is the discrete forward operator acting on $\mathbf{x}_k^t$ and $\mathbf{e}^{obs} \equiv \mathbf{e}_k^r + \mathbf{e}_k^m$ is the total observation error. The measurement error is $\mathbf{e}_k^m$ while the representativeness error (see Lorenc [2]) $\mathbf{e}_k^r$ is given by the difference between the representation of the continuum forward model and formulating the model error discretely, that is:

$$\mathbf{e}_k^r = \mathbf{e}_k^r(\mathbf{x}_k) \equiv \mathbf{h}_k^c(\mathbf{x}_k) - \mathbf{h}_k(\mathbf{\Pi x}_k). \tag{2.10}$$

The impact of this error on our system will be more clear when we address the initialization problem in the next paragraph.

2.1.1 The Estimation Problem

Since our goal is to study a physical system described by the vector state $\mathbf{x}_k^t$ and since this vector is unknown, one assumes that the best estimation of the system is given by the state vector $\mathbf{x}_k^b$, where b stands for background, denoting the a priori or background estimate of the true state before the analysis is carried out, valid at the same time. This vector is the result of the data assimilation or statistical analysis performed earlier. By Eq. 2.9 the observations performed on the system bring new information through the operator $\mathbf{h}_k$. One assumes the statistics of the observation error are known up to the second order moments. When new observations are available we can improve the analysis obtaining our estimation $\mathbf{x}_k^a$ with its error. The suffix a means analysis.

In geophysics it is usual to define the a priori estimate as forecast/background and the posterior estimate as analysis.

Let us now explore the background and analysis errors. The background error is defined as:

$$\mathbf{e}_k^b = \mathbf{x}_k^b - \mathbf{x}_k^t \tag{2.11}$$

it reflects the discrepancy between the a priori estimate and the unknown truth. It is considered stochastic the mean of which is $\hat{\mathbf{e}}^b = E[\mathbf{e}_k^b]$. The background error covariance is: $\mathbf{B} = E[(\mathbf{e}_k^b - \hat{\mathbf{e}}_k^b)(\mathbf{e}_k^b - \hat{\mathbf{e}}_k^b)^T]$. The analysis error is defined as:

$$\mathbf{e}_k^a = \mathbf{x}_k^a - \mathbf{x}_k^t \tag{2.12}$$

It defines the difference between the analysis process and the truth; the related analysis error covariance is: $\mathbf{A} = E[(\mathbf{e}_k^a - \hat{\mathbf{e}}_k^a)(\mathbf{e}_k^a - \hat{\mathbf{e}}_k^a)^T]$ with its mean given by: $\hat{\mathbf{e}}^a = E[\mathbf{e}_k^a]$. All matrices are symmetric and positive.

2.1.2 The Linear Hypothesis

In order to understand if we can linearize our system we need to analyze the role played by the representativeness error in the model. If we write the representativeness error adding and subtracting $\mathbf{h}_k^c(\boldsymbol{\Pi}\mathbf{x}_k)$ to the relation (2.10)

$$\mathbf{e}_k^r = \mathbf{h}_k^c(\mathbf{x}_k) - \mathbf{h}_k(\boldsymbol{\Pi}\mathbf{x}_k) + \mathbf{h}_k^c(\boldsymbol{\Pi}\mathbf{x}_k) - \mathbf{h}_k^c(\boldsymbol{\Pi}\mathbf{x}_k), \tag{2.13}$$

we can split the representativeness error into the sum of two parts, $\mathbf{e}_k^{'}$ and $\mathbf{e}_k^{''}$ giving:

$$\begin{aligned} \mathbf{e}_k^{'} &\equiv \mathbf{h}_k^c(\mathbf{x}_k) - \mathbf{h}_k^c(\boldsymbol{\Pi}\mathbf{x}_k) \\ \mathbf{e}_k^{''} &\equiv \mathbf{h}_k^c(\boldsymbol{\Pi}\mathbf{x}_k) - \mathbf{h}_k(\boldsymbol{\Pi}\mathbf{x}_k), \end{aligned} \tag{2.14}$$

where $\mathbf{e}_k^{''}$ can be all depending on our integration interpolation formulas and is easy to solve, while $\mathbf{e}_k^{'}$ depending on the scale variability of $\mathbf{x}_k$ could dominate the measurement error, as happens in a highly variable field. This can be more clarified if we use a linear approximation of $\mathbf{e}_k^{'}$, for instance applying the first order Taylor expansion to $\mathbf{h}_k^c$ obtaining:

$$\mathbf{e}_k^{'} = \mathbf{H}_k^c(\mathbf{I} - \boldsymbol{\Pi})\mathbf{x}_k \tag{2.15}$$

where $\mathbf{H}_k^c$ is:

$$\mathbf{H}_k^c = \left.\frac{\partial \mathbf{h}_k^c(\mathbf{x}_k)}{\partial \mathbf{x}}\right|_{\mathbf{x}=\boldsymbol{\Pi}\mathbf{x}_\mathbf{k}}, \tag{2.16}$$

that operates on the unresolved portion of $(\mathbf{I} - \boldsymbol{\Pi})\mathbf{x}_k$ of the continuum state $\mathbf{x}_k$. $\mathbf{H}_k^c$ is a tangent linear operator or Jacobian matrix of a non linear operator. If we define the linear operator as $\mathbf{H}$ we can write $\mathbf{H}_k^c = \mathbf{H}$

Thus the vector of observations $\mathbf{y}$ is related with the observation operator $\mathbf{H}(\mathbf{x})$ and the observation error as $\mathbf{e}$ by:

$$\mathbf{y} = \mathbf{H}\mathbf{x}^t + \mathbf{e}. \tag{2.17}$$

Let us estimate now, with a reasonable assumption, that $\mathbf{x}^a$ is a linear combination of all available information given by the $n \times n$ matrix $\mathbf{L}$ and $n \times p$ matrix where $\mathbf{K}$ is the linear operator, we have:

$$\mathbf{x}^a = \mathbf{L}\mathbf{x}^b + \mathbf{K}\mathbf{y} \tag{2.18}$$

Given the observation Eq. 2.17 we obtain that the estimate error of the analysis is:

$$\begin{aligned} \mathbf{x}^a - \mathbf{x}^t &= \mathbf{L}(\mathbf{x}^b - \mathbf{x}^t + \mathbf{x}^t) + \mathbf{K}(\mathbf{H}\mathbf{x}^t + \mathbf{e}) - \mathbf{x}^t \\ \mathbf{e}^a &= \mathbf{L}\mathbf{e}^b + \mathbf{K}\mathbf{e} + (\mathbf{L} + \mathbf{K}\mathbf{H} - \mathbf{I})\mathbf{x}^t. \end{aligned} \tag{2.19}$$

If we assume the errors and observations are unbiased, the related expectations are $E[\mathbf{e}] = 0$ and $E[\mathbf{e}^b] = 0$ and thus $E[\mathbf{e}^a] = (\mathbf{L} + \mathbf{KH} - \mathbf{I})E[\mathbf{x}^t]$. On the contrary if there is a bias, it is always possible to diagnose it and subtract its value from the total observation errors to make the corrected error unbiased. If we postulate that:

$$\mathbf{L} = \mathbf{I} - \mathbf{KH}, \tag{2.20}$$

as sufficient but not necessary condition, we have:

$$\begin{aligned} \mathbf{x}^a &= (\mathbf{I} - \mathbf{KH})\mathbf{x}^b + \mathbf{Ky} \\ \mathbf{x}^a &= \mathbf{x}^b + \mathbf{K}(\mathbf{y} - \mathbf{Hx}^b), \end{aligned} \tag{2.21}$$

where the difference $[\mathbf{y} - \mathbf{Hx}^b]$ is called innovation and $\mathbf{K}$ is a weight or gain computed on the estimated statistical error covariances of the forecast and the observations. It gives the additional information brought in by the observation compared to background. The error covariance matrix obtained from the innovation vector is called information matrix. Since the operator $\mathbf{K}$ is linear, the analysis is a linear interpolation, as was for example in the first Cressman [3] interpolation.

Once we have computed the optimal gain matrix $\mathbf{K}$ we need to compute the error covariance or posterior error matrix $\mathbf{P}^a$.

Using the notation already introduced for the error we have:

$$\mathbf{e}^a = \mathbf{e}^b + \mathbf{K}(\mathbf{e} - \mathbf{He}^b) \tag{2.22}$$

As previously we assume the errors and observations are unbiased so that the covariance analysis matrix is $\mathbf{P}^a = E[(\mathbf{e}^a)(\mathbf{e}^a)^T]$. Remembering the error covariance background matrix $\mathbf{P}^b$ and error covariance observation matrix $\mathbf{R}$ and the relation (2.20) we have:

$$\begin{aligned} \mathbf{P}^a &= E[(\mathbf{e}^a)(\mathbf{e}^a)^T] \\ &= E[(\mathbf{e}^b + \mathbf{K}(\mathbf{e} - \mathbf{He}^b))(\mathbf{e}^b + \mathbf{K}(\mathbf{e} - \mathbf{He}^b))^T] \\ &= E[(\mathbf{Le}^b + \mathbf{Ke})(\mathbf{Le}^b + \mathbf{Ke})^T] \\ &= E[\mathbf{Le}^b(\mathbf{e}^b)^T\mathbf{L}^T] + E[\mathbf{Ke}(\mathbf{e})^T\mathbf{K}^T] \\ &= \mathbf{LP}^b\mathbf{L}^T + \mathbf{KRK}^T \\ &= (\mathbf{I} - \mathbf{KH})\mathbf{P}^b(\mathbf{I} - \mathbf{KH})^T + \mathbf{KRK}^T \end{aligned} \tag{2.23}$$

that is the posterior error.

In order to obtain an optimal estimation, we minimize the trace of the analysis error covariance given by (2.23). Remembering $Trace[\mathbf{P}^b] = Trace[\mathbf{P}^b]^T$ and $Trace[\mathbf{R}] = Trace[\mathbf{R}]^T$, expanding the relation (2.23) we have:

$$
\begin{aligned}
Trace[\mathbf{P}^a] =& Trace[\mathbf{P}^b] + Trace[\mathbf{KHP}^b\mathbf{H}^T\mathbf{K}^T] - 2Trace[\mathbf{P}^b\mathbf{H}^T\mathbf{K}^T] \\
&+ Trace[\mathbf{KRK}^T]
\end{aligned}
\tag{2.24}
$$

Following Bouttier and Courtier [4] we have a continuous differentiable scalar function of the coefficient of $\mathbf{K}$ whose first order derivative in $\mathbf{K}$ of the difference $Trace[\mathbf{P}^a](\mathbf{K}+\mathbf{L}) - Trace[\mathbf{P}^a](\mathbf{K})$, where $\mathbf{L}$ is an arbitrary test matrix, is:

$$
\begin{aligned}
\frac{d[Trace[\mathbf{P}^a]]\mathbf{L}}{d\mathbf{K}} &= 2Trace[\mathbf{KHP}^b\mathbf{H}^T\mathbf{L}^T] - 2Trace[\mathbf{P}^b\mathbf{H}^T\mathbf{L}^T] + 2Trace[\mathbf{KRL}^T] \\
&= 2Trace[\mathbf{KHP}^b\mathbf{H}^T\mathbf{L}^T - \mathbf{P}^b\mathbf{H}^T\mathbf{L}^T + \mathbf{KRL}^T] \\
&= 2Trace\{[\mathbf{K}(\mathbf{HP}^b\mathbf{H}^T + \mathbf{R}) - \mathbf{P}^b\mathbf{H}^T]\mathbf{L}^T\}
\end{aligned}
\tag{2.25}
$$

The last line shows that the derivative is zero for any choice of $\mathbf{L}$ if $(\mathbf{HP}^b\mathbf{H}^T + \mathbf{R})\mathbf{K}^T - \mathbf{P}^b\mathbf{H} = 0$ that is equivalent to:

$$
\mathbf{K} = \mathbf{P}^b\mathbf{H}^T(\mathbf{HP}^b\mathbf{H}^T + \mathbf{R})^{-1} \tag{2.26}
$$

because $(\mathbf{HP}^b\mathbf{H}^T + \mathbf{R})$ is assumed invertible.

The estimate of $\mathbf{x}^a$ and $\mathbf{P}^a$ are called **Best Linear Unbiased Estimator, BLUE**. **BLUE**, because $\mathbf{H}$ is linear through $\mathbf{K}$ and $\mathbf{L}$, without bias through the first step of derivation and optimal through the second derivation.

2.1.3 Optimal Estimation

Now we need to improve our knowledge of the state $\mathbf{x}^a$ taking into account the two available sources of information: the model and observations. There are two ways to combine observations with the model:

- the observations $\mathbf{y}$ may be interpolated between the observational data which can be sparse in time and space subjected to constraints provided by the model;
- we want to reduce the uncertainties of model $\mathbf{H}$ on the input $\mathbf{x}$, under the constrains of the measurements.

We define a cost function $\mathcal{J}$ that is a measure of the distance between the observations and model.

$$
\mathcal{J}(\mathbf{x}) = ||\mathbf{y} - \mathbf{H}(\mathbf{x})||^2, \tag{2.27}
$$

with $||\cdot||$ the norm two. Since we need to balance each component through the confidence in the measurement we can access to the a priori estimate or background information. In this way we introduce a compromise between the observations given by the observation and the information given by background value. Then the cost function can be defined as

$$
\mathcal{J}(\mathbf{x}) = \alpha \times ||\mathbf{y} - \mathbf{H}(\mathbf{x})||^2 + \beta \times ||\mathbf{x} - \mathbf{x}^b||^2. \tag{2.28}
$$

α and β are the weight given to the confidence in the observations and background. Those parameters can be defined empirically or analytically, knowing the background and observation errors.

A simple example drawn by Bouttier and Courtier [4] based on temperature and related error variance shows that the cost function terms which have a quadratic form tend to pull the analysis $\mathbf{x}^a$ toward the background $\mathbf{x}^b$ and the observation $\mathbf{y}$, respectively. In this case $\mathbf{x}^a$ makes $\mathcal{J}(\mathbf{x})$ as small as possible, given the computational constraints.

The quadratic form (2.28) can be written in matrix form as:

$$\mathcal{J}(\mathbf{x}) = \frac{1}{2}\{(\mathbf{y} - \mathbf{H}(\mathbf{x}))^T \mathbf{R}^{-1}(\mathbf{y} - \mathbf{H}(\mathbf{x})) + (\mathbf{x} - \mathbf{x}^b)^T (\mathbf{P}^b)^{-1}(\mathbf{x} - \mathbf{x}^b)\}, \quad (2.29)$$

where $\mathbf{y}$ is the vector of observation of length k; $\mathbf{x}^b$ is a vector of background of length j and $\mathbf{P}^b$ is the background error covariance matrix of rank $j \times j$ and $\mathbf{R}$ is the observation error covariance matrix of rank $k \times k$. $\mathbf{H}$ is the linear forward interpolation operator.

The optimal solution, i.e. the analysis $\mathbf{x}^a$ that is closest to the *true* state $\mathbf{x}^t$, in an *r.m.s.* sense, requires the minimum:

$$\mathcal{J}(\mathbf{x}) \overset{x}{\rightarrow} min \quad (2.30)$$

It demands the first derivative, the gradient, of the cost function with respect to its variable $\mathbf{x}$ at the analysis $\mathbf{x}^a$ be equal zero

$$\nabla\mathcal{J}(\mathbf{x}^a) = \left.\frac{d\mathcal{J}}{d\mathbf{x}}\right|_{\mathbf{x}^a} = 0 \quad (2.31)$$

Under our linear assumption, we have:

$$(\mathbf{P}^b)^{-1}(\mathbf{x}^a - \mathbf{x}^b) - \mathbf{H}^T\mathbf{R}^{-1}[\mathbf{y} - \mathbf{H}(\mathbf{x}^a)] = 0 \quad (2.32)$$

Adding and subtracting $\mathbf{y} - \mathbf{H}(\mathbf{x}^b)$ we obtain:

$$(\mathbf{P}^b)^{-1}(\mathbf{x}^a - \mathbf{x}^b) - \mathbf{H}^T\mathbf{R}^{-1}[\mathbf{y} - \mathbf{H}(\mathbf{x}^b)] - \mathbf{H}^T\mathbf{R}^{-1}[\mathbf{H}(\mathbf{x}^a) - \mathbf{H}(\mathbf{x}^b)] = 0 \quad (2.33)$$

Rearranging this equation and applying again the linear assumption, we have:

$$0 = (\mathbf{P}^b)^{-1}(\mathbf{x}^a - \mathbf{x}^b) - \mathbf{H}^T\mathbf{R}^{-1}[\mathbf{y} - \mathbf{H}(\mathbf{x}^b)] - \mathbf{H}^T\mathbf{R}^{-1}\mathbf{H}(\mathbf{x}^a - \mathbf{x}^b)$$
$$\mathbf{x}^a - \mathbf{x}^b = ((\mathbf{P}^b)^{-1} + \mathbf{H}^T\mathbf{R}^{-1}\mathbf{H})^{-1}\mathbf{H}^T\mathbf{R}^{-1}[\mathbf{y} - \mathbf{H}(\mathbf{x}^b)] \quad (2.34)$$

The analysis state $\mathbf{x}^a$ is called optimal because is closest in a root mean square sense to the *true* state $\mathbf{x}^t$. The equivalence of this relation with relation (2.21) can also be shown applying the Sherman-Woodbury-Morrinson equation.[2] In fact, remembering that $\mathbf{K}$ is given by the relation (2.26) we have:

$$\begin{aligned}
\mathbf{K} &= \mathbf{P}^b\mathbf{H}^T(\mathbf{H}\mathbf{P}^b\mathbf{H}^T+\mathbf{R})^{-1}\\
&= ((\mathbf{P}^b)^{-1}+\mathbf{H}^T\mathbf{R}^{-1}\mathbf{H})^{-1}((\mathbf{P}^b)^{-1}+\mathbf{H}^T\mathbf{R}^{-1}\mathbf{H})\mathbf{P}^b\mathbf{H}^T(\mathbf{H}\mathbf{P}^b\mathbf{H}^T+\mathbf{R})^{-1}\\
&= ((\mathbf{P}^b)^{-1}+\mathbf{H}^T\mathbf{R}^{-1}\mathbf{H})^{-1}(\mathbf{H}^T+\mathbf{H}^T\mathbf{R}^{-1}\mathbf{H}\mathbf{P}^b\mathbf{H}^T)(\mathbf{H}\mathbf{P}^b\mathbf{H}^T+\mathbf{R})^{-1}\\
&= ((\mathbf{P}^b)^{-1}+\mathbf{H}^T\mathbf{R}^{-1}\mathbf{H})^{-1}\mathbf{H}^T\mathbf{R}^{-1}(\mathbf{R}+\mathbf{H}\mathbf{P}^b\mathbf{H}^T)(\mathbf{H}\mathbf{P}^b\mathbf{H}^T+\mathbf{R})^{-1}\\
&= ((\mathbf{P}^b)^{-1}+\mathbf{H}^T\mathbf{R}^{-1}\mathbf{H})^{-1}\mathbf{H}^T\mathbf{R}^{-1}.
\end{aligned} \tag{2.36}$$

The equivalence can be useful because sometime the inversion of the $((\mathbf{P}^b)^{-1}+\mathbf{H}^T\mathbf{R}^{-1}\mathbf{H})^{-1}$ is more costly than the matrix $\mathbf{R}+\mathbf{H}\mathbf{P}^b\mathbf{H}^T$.

Summarizing we have:

$$\begin{aligned}
\mathbf{x}^a &= \mathbf{x}^b+\mathbf{K}[\mathbf{y}-\mathbf{H}(\mathbf{x}^b)]\\
\mathbf{W} &= \mathbf{P}^b\mathbf{H}^T[\mathbf{R}+\mathbf{H}\mathbf{P}^b\mathbf{H}^T]^{-1}.
\end{aligned} \tag{2.37}$$

With respect to the relaxation method Optimal Interpolation is an intermittent assimilation methods. It is used at synoptic times, that is the instant of standard meteorological time. The difference of $\mathbf{K}$ gain with respect to the Kalman gain is that the background error covariance matrix $\mathbf{P}_k^b$ is specified rather than predicted. In optimal estimation the element of $\mathbf{P}_k^b$ is based on statistical evaluations and dynamic constraints. For the state vector $\mathbf{r}(\lambda, \psi, \phi)$ at a certain instant, the error covariance matrix $\mathbf{P}^b$ is:

$$\mathbf{P}^b = E[\mathbf{e}(\mathbf{r}_i)\mathbf{e}^T(\mathbf{r}_j)] \tag{2.38}$$

where $\mathbf{e}(\mathbf{r}) = \mathbf{x}^b(\mathbf{r}) - \mathbf{x}^t(\mathbf{r})$ is the forecast error with $\mathbf{x}^t(\mathbf{r})$ represents the true value of the state vector and $\mathbf{x}^b(\mathbf{r})$ is the forecast state vector. Therefore we can decompose $\mathbf{P}^b$ as

$$\mathbf{P}^b(\mathbf{r}_i,\mathbf{r}_j) \equiv \begin{vmatrix} \mathbf{P}^{b|\lambda\lambda} & \mathbf{P}^{b|\lambda\psi} & \mathbf{P}^{b|\lambda\phi}\\ \mathbf{P}^{b|\psi\lambda} & \mathbf{P}^{b|\psi\psi} & \mathbf{P}^{b|\psi\phi}\\ \mathbf{P}^{b|\phi\lambda} & \mathbf{P}^{b|\phi\psi} & \mathbf{P}^{b|\phi\phi} \end{vmatrix}$$

where $\mathbf{P}^{b|\cdot\cdot}$ are the cross-covariance functions defined in analogy to (2.38).

It is a simplification of the algebraic calculation of the gain matrix. The first of the Eq. (2.37) is solved through the inversion, the matrix $\mathbf{K}$ is simplified assuming that only the most forthcoming observations determine the analysis increment. For each of the variables of the model the increment of the analysis is given from the corresponding $\mathbf{K}$ for the vector of deviations of the background value $[\mathbf{y}-\mathbf{H}(\mathbf{x}^b)]$.

[2] Sherman-Morrison-Woodbury formula is:

$$(\mathbf{A}+\mathbf{UCV})^{-1} = \mathbf{A}^{-1} - \mathbf{A}^{-1}\mathbf{V}(\mathbf{C}^{-1}+\mathbf{V}\mathbf{A}^{-1}\mathbf{V})^{-1}\mathbf{V}\mathbf{A}^{-1}, \tag{2.35}$$

where all matrices have they correct size.

Bouttier, and Courtier [4] provide the fundamental assumptions and procedures to be followed.

The fundamental hypothesis of optimal interpolation (OI) is that: for each variable of the model, just some observations are important to determine the increase of the analysis. From this it follows that:

1. for each variable of the model $\mathbf{x}(i)$ choose a small number of observations p_i using an empirical policy of selection;
2. form the corresponding p_i list of deviations of the background data $[\mathbf{y} - \mathbf{H}(\mathbf{x}^b)]_i$, of the errors background covariance matrix, between the variables of the model $x(i)$ and the state of the model interpolated in points p_i namely the p_i coefficients of the ith row of $\mathbf{P}^b\mathbf{H}$ and $(p_i \times p_i)$ covariance submatrix of the errors observations and background formed by $\mathbf{HP}^b\mathbf{H}^T$ and $\mathbf{R}$ for selected observations;
3. invert the positive-definite matrix $(p_i \times p_i)$ formed from $[\mathbf{R} + \mathbf{HP}^b\mathbf{H}^T]$ for selected observations (e.g. using Cholesky Factorizazion methods or LU);
4. multiply it by the ith row of $\mathbf{P}^b\mathbf{H}$ to get the $\mathbf{K}$ row required.

In Optimal Interpolation it is necessary that $\mathbf{P}^b$ is a matrix that can be easily applied to a pair of observed value and model variables or to a pair of observed variables. The simplicity of the OI collides with the disadvantage that there is no consistency between small and large scales and that $\mathbf{H}$ must be linear.

2.1.4 Minimization Methods of Cost Functions

There are several methods to minimize the cost function including the most relevant below. If the cost function is quadratic and convex, its solution is unique. In general, however, $\mathcal{J}$ exhibits several minima. In such a case the problem is more difficult to solve, even though there are some algorithms among which we select conjugate gradient and quasi-Newton methods.

The minimization algorithms start from an initial point $\mathbf{x}_0$ and construct a sequence $\mathbf{x}_k$ which converges to a local minimum. At each step k one determines a direction $\mathbf{d}_k$ to define the next point of the sequence. Then the problem of minimization of multivariable functions is usually solved by determining a search direction vector $\mathbf{d}_k$ and solve it as a linear minimization problem. If $\mathbf{x}_k$ is the vector containing the variables to be determined and $\mathbf{d}_k$ is the vector of search direction, at each iteration step the minimization problem of a function $\mathbf{f}$ is formulated so as to find the step size λ that minimizes $\mathbf{f}(\mathbf{x} + \lambda\mathbf{d}_k)$, where λ is a positive and real number. At the next iteration, $\mathbf{x}$ is replaced by $\mathbf{x}_k + \lambda_k\mathbf{d}_k$ and a new search direction is determined.

The conjugate gradient method is an algorithm for finding the nearest local minimum of a function which uses conjugate directions for descending. Two vectors $\mathbf{u}$ and $\mathbf{v}$ are said to be conjugate, with respect to a matrix $\mathbf{A}$, if

$$\mathbf{u}^T\mathbf{A}\mathbf{v} = 0, \tag{2.39}$$

where $\mathbf{A}$ is the Hessian matrix of the cost function. In Press et al's book [5] there are two conjugate gradient methods by Fletcher-Reeves and Polak-Ribière.

These algorithms calculate the mutually conjugate directions of search with respect to the Hessian matrix of the cost function directly from the function and the gradient evaluations, but without the direct evaluation of the Hessian matrix. The new search direction $\mathbf{d}_{k+1}$ is determined by using

$$\mathbf{d}_{k+1} = -\mathbf{g}_{k+1} + \lambda_k \mathbf{d}_k, \tag{2.40}$$

where $\mathbf{d}_k$ is the previous search direction, $\mathbf{g}_{k+1}$ is the local gradient at iteration step $k + 1$ that is determined by the Fletcher-Reeves equation

$$\lambda_k = \frac{\mathbf{g}_{k+1} \cdot \mathbf{g}_{k+1}}{\mathbf{g}_k \cdot \mathbf{g}_k} \tag{2.41}$$

and the Polak-Ribière equation

$$\lambda_k = \frac{(\mathbf{g}_{k+1} - \mathbf{g}_k) \cdot \mathbf{g}_{k+1}}{\mathbf{g}_k \cdot \mathbf{g}_k}. \tag{2.42}$$

If the vicinity of the minimum has the shape of a long, narrow valley, the minimum is reached in far fewer steps than would be the case using the steepest descent method, which makes use of the inverse of the local gradient as the search direction. The line minimization to find the step size λ that minimizes $\mathbf{f}(\mathbf{x} + \lambda \mathbf{d}_k)$ at every iteration step can be done by using the Golden Section Search Algorithm [5].

For the problem of minimizing a multivariable function quasi-Newton methods are also widely used. These methods involve the approximation of the Hessian, or its inverse, matrix of the function. The LBFGS (Limited memory-Broyden-Fletcher-Goldfarb-Shanno) method is basically a method to approximate the Hessian matrix in the quasi-Newton method of optimization. It is a variation of the standard BFGS method, which is given by (Nocedal [6], Byrd et al. [7])

$$\mathbf{x}_{k+1} = \mathbf{x}_k - \lambda_k \mathcal{H}_k \mathbf{g}_k, \quad k = 1, 2, 3 \ldots \tag{2.43}$$

where λ_k is a step length, $\mathbf{g}_k$ is the local gradient of the cost function, and $\mathcal{H}_k$ is the approximate inverse Hessian matrix which is updated at every iteration by means of the formula

$$\mathcal{H}_{k+1} = \mathbf{V}_k^T \mathcal{H}_k \mathbf{V}_k + \rho_k \mathbf{s}_k \mathbf{s}_k^T \tag{2.44}$$

where

$$\rho_k = \frac{1}{\mathbf{q}^T \mathbf{s}_k} \tag{2.45}$$

and

$$\mathbf{V}_k = \mathbf{I} - \rho_k \mathbf{q} \mathbf{s}_k^T \tag{2.46}$$

$$\mathbf{s}_k = \mathbf{s}_{k+1} - \mathbf{s}_k \tag{2.47}$$

and

$$\mathbf{q}_k = \mathbf{g}_{k+1} - \mathbf{g}_k \tag{2.48}$$

Using this method, instead of storing the matrices $\mathcal{H}_k$, one stores a certain number of pairs $\{\mathbf{s}_k, \mathbf{q}_i\}$ that define them implicitly. The product of $\mathcal{H}_k \mathbf{g}_k$ is obtained by performing a sequence of inner products involving $\mathbf{g}_k$ and the most recent vector pairs $\{\mathbf{s}_k, \mathbf{q}_i\}$ to define the iteration matrix.

2.1.5 Some Properties of Estimation

1. The innovation and the analysis residue $\mathbf{y} - \mathbf{H}\mathbf{x}^a$ are unbiased.
   As consequence of the linearity of the observation operator and absence of bias of the analysis error $\mathbf{e}^a$ the analysis residue is unbiased. In fact:

$$\mathbf{y} - \mathbf{H}\mathbf{x}^a = \mathbf{H}\mathbf{x}^t + \mathbf{e} - \mathbf{H}\mathbf{x}^a = \mathbf{H}\mathbf{e}^a - \mathbf{e} \tag{2.49}$$

   and we can conclude $E[\mathbf{y} - \mathbf{H}\mathbf{x}^a] = 0$. Since the background error is unbiased also the innovation error is also unbiased

$$\mathbf{y} - \mathbf{H}\mathbf{x}^b = \mathbf{H}\mathbf{x}^t + \mathbf{e} - \mathbf{H}\mathbf{x}^b = \mathbf{H}\mathbf{e}^b - \mathbf{e}, \tag{2.50}$$

   from which $E[\mathbf{y} - \mathbf{H}\mathbf{x}^b] = 0$.
2. The analysis and analysis error are orthogonal.
   Calculate the covariance matrix

$$\mathbf{C} = E[\mathbf{x}^a (\mathbf{e}^a)^T]. \tag{2.51}$$

   Assuming the background satisfies

$$E[\mathbf{x}^b (\mathbf{e}^b)^T] = 0 \tag{2.52}$$

   means the background and its error are uncorrelated. From Eq. 2.21 we have

$$\mathbf{x}^a = \mathbf{x}^b + \mathbf{K}(\mathbf{y} - \mathbf{H}\mathbf{x}^b) = \mathbf{x}^b + \mathbf{K}(-\mathbf{H}\mathbf{e}^b - \mathbf{e}). \tag{2.53}$$

Then, recalling (2.22)

$$\begin{aligned}\mathbf{C} &= E[(\mathbf{x}^b + \mathbf{K}(-\mathbf{H}\mathbf{e}^b - \mathbf{e}))((\mathbf{I} - \mathbf{K}\mathbf{H})\mathbf{e}^b - \mathbf{K}\mathbf{e}^T] \\ &= -\mathbf{K}\mathbf{H}\, E[\mathbf{e}^b(\mathbf{e}^b)^T](\mathbf{I} - \mathbf{K}\mathbf{H})^T + \mathbf{K}\, E[\mathbf{e}\mathbf{e}^T]\mathbf{K}^T \\ &= \mathbf{K}[\mathbf{H}\mathbf{P}^b(-\mathbf{I} - \mathbf{K}\mathbf{H})^T + \mathbf{R}\mathbf{K}^T]. \end{aligned} \quad (2.54)$$

If the analysis is optimal, we have $-\mathbf{H}\mathbf{P}^b(\mathbf{I} - \mathbf{K}\mathbf{H})^T + \mathbf{R}\mathbf{K}^T = 0$, so $\mathbf{C} = 0$ and the estimate $\mathbf{x}^a$ and its error $\mathbf{e}^a$ are orthogonal.

2.1.6 Estimation of the quality of analysis

An important step in the process of assimilation is to be able to estimate the quality of the analysis. In fact in a sequential analysis it is useful to know the level of reliability of the analysis because it helps to specify the background error covariance matrix for later analysis. If the background is a forecast then as we have seen the errors are a combination of the errors of the model and those of the analysis that evolve in time according to dynamic model as it is also seen using of the algorithm of Kalman filter.

The word quality means reliability and it is estimated through the value of the covariance matrix of the error of the analysis $\mathbf{P}^a$. The process by which we estimate the quality of the analysis is closely linked to the cost or penalty function and its gradient. Recalling the relation (2.29) and (2.31), the second derivative or Hessian of the cost function derived two times around x that the control variable is:

$$\nabla\nabla\mathcal{J}(x) = 2((\mathbf{P}^b)^{-1} + \mathbf{H}^T\mathbf{R}^{-1}\mathbf{H}) \quad (2.55)$$

Introducing the true state x^t of the model into the (2.31) we have:

$$\begin{aligned} 0 &= (\mathbf{P}^b)^{-1}(\mathbf{x}^a - \mathbf{x}^t + \mathbf{x}^t - \mathbf{x}^a) - \mathbf{H}^T\mathbf{R}^{-1}(\mathbf{y} - \mathbf{H}(\mathbf{x}^t) + \mathbf{H}(\mathbf{x}^t) - \mathbf{x}^a) \\ &= (\mathbf{P}^b)^{-1}(\mathbf{x}^a - \mathbf{x}^t) - \mathbf{H}^T\mathbf{R}^{-1}\mathbf{H}(\mathbf{x}^t - \mathbf{x}^a) - (\mathbf{P}^b)^{-1}(\mathbf{x}^b - \mathbf{x}^t) + \mathbf{H}^T\mathbf{R}^{-1}(\mathbf{y} - \mathbf{H}(\mathbf{x}^t)). \end{aligned} \quad (2.56)$$

Thus

$$((\mathbf{P}^b)^{-1} + \mathbf{H}^T\mathbf{R}^{-1}\mathbf{H})(\mathbf{x}^a - \mathbf{x}^t) = (\mathbf{P}^b)^{-1}(\mathbf{x}^b - \mathbf{x}^t) + \mathbf{H}^T\mathbf{R}^{-1}(\mathbf{y} - \mathbf{H}(\mathbf{x}^t)). \quad (2.57)$$

Multiplying the right side of this equation by its transposed and taking into account of the Eq. 2.23 and computing the expectation we have:

$$\begin{aligned} &((\mathbf{P}^b)^{-1} + \mathbf{H}^T\mathbf{R}^{-1}\mathbf{H})\mathbf{P}^a((\mathbf{P}^b)^{-1} + \mathbf{H}^T\mathbf{R}^{-1}\mathbf{H})^T \\ &\quad = (\mathbf{P}^b)^{-1}\mathbf{P}^b(\mathbf{P}^b)^{-1} + \mathbf{H}^T\mathbf{R}^{-1}\mathbf{R}\mathbf{R}^{-1}\mathbf{H} \quad (2.58) \\ &+[(\mathbf{P}^b)^{-1}\mathbf{H}^T\mathbf{R}^{-1}\overline{(\mathbf{x}^b - \mathbf{x}^t)^T(\mathbf{y} - \mathbf{H}(\mathbf{x}^t)} + (\mathbf{P}^b)^{-1}\mathbf{R}^{-1}\overline{(\mathbf{x}^b - \mathbf{x}^t)(\mathbf{y} - \mathbf{H}(\mathbf{x}^t)^T}. \quad (2.59) \end{aligned}$$

Assuming the background error and those due to the observation are uncorrelated, by simplifying we obtain:

$$((\mathbf{P}^b)^{-1} + \mathbf{H}^T\mathbf{R}^{-1}\mathbf{H})\mathbf{P}^a((\mathbf{P}^b)^{-1} + \mathbf{H}^T\mathbf{R}^{-1}\mathbf{H})^T = ((\mathbf{P}^b)^{-1} + \mathbf{H}^T\mathbf{R}^{-1}\mathbf{H}). \quad (2.60)$$

Removing the zero component we have:

$$\mathbf{P}^a = ((\mathbf{P}^b)^{-1} + \mathbf{H}^T\mathbf{R}^{-1}\mathbf{H})^{-1} \quad (2.61)$$

By definition, the Hessian is given by (2.55) and:

$$\mathbf{P}^a = \frac{1}{2}[\nabla\nabla\mathcal{J}(x)]^{-1} \quad (2.62)$$

or inversely

$$\nabla\nabla\mathcal{J}(x) = \frac{1}{2}(\mathbf{P}^a)^{-1} \quad (2.63)$$

The matrix $(\mathbf{P}^a)^{-1}$ is called information matrix.

2.2 Variational Approach: 3-D VAR and 4-D VAR

The basic principle of the $3D - Var$ is to avoid explicitly calculating the gain matrix and make its inversion using a minimization procedure of the cost function $\mathcal{J}$. In this case the solution of the Eq. (2.29) is obtained iteratively doing various evaluations of the equation and its gradient to get the minimum using a suited descent algorithm. The minimization is obtained limiting the number of iterations and stipulating that the norm of the gradient $\| \nabla\mathcal{J}(s^a) \|$ decreases by a predefined amount, during minimization this is an intrinsic measure of how close the analysis is to the optimal value rather than the starting point of minimization.

When the observations are distributed over time the approach $3D - Var$ is generalized to the approach $4D - Var$. The equations are the same provided the operators are generalized including a forecasting model that allows comparison the state of the model with the observations at a time k defined.

In a given time interval the cost function to be minimized is the same of the $3D - Var$ but with a difference related to the operator H and R that are subject to a partial trajectory, i.e. the k ranges from $k - 1$ to k. Thus according to the relation given by Lorenc [2]:

$$\mathcal{J} = \frac{1}{2}\left\{\sum_{t=0}^{N}[\mathbf{y}_k - \mathbf{H}_k(\mathbf{x}_t)]^T\mathbf{R}_k^{-1}[\mathbf{y}_k - \mathbf{H}_k(\mathbf{x}_k)]\right\} + \frac{1}{2}[\tilde{\mathbf{x}} - \mathbf{x}_0^b]^T(\mathbf{P}^b)^{-1}[\tilde{\mathbf{x}} - \mathbf{x}_0^b], \quad (2.64)$$

where $\tilde{\mathbf{x}}$ is at zero time forecast produced by the data assimilation for $k < k_0$. $\mathbf{P}^b$ is the covariance matrix of the errors while $k = 0$; $\mathbf{x}_0^b$ is a value of $\tilde{\mathbf{x}}$ before the first iteration of the descent algorithm. The second term on the right it is thought to force the forecast $\mathbf{x}_0^b$ towards the prediction before $\tilde{\mathbf{x}}$ to reduce the time discontinuity. In the classical receipt (see [8]) the assimilation issue $4D - Var$ is subject to a strong constraint such that the sequence of model states $\mathbf{x}_t$ must be a solution of the equation:

$$\mathbf{x}_k = \mathbf{M}_{0\to k}(\mathbf{x}) \quad \forall k \tag{2.65}$$

where $\mathbf{M}_{0\to k}$ is a forecasting model from the starting time to k. $4D - Var$ is a problem of nonlinear optimization difficult to solve unless in the following two hypotheses: causality and tangent linear.

1. Causality
   The forecast model can be expressed as the product of intermediate forecasting steps, that reflect the randomness of nature. The integration of a prognostic model starts with the initial condition $\mathbf{x}_0 = \mathbf{x}$ so that $\mathbf{M}_0$ is the identity. Thus indicating with $\mathbf{M}_k$ the step of forecasting from $k - 1$ to k we have $\mathbf{x}_t = \mathbf{M}_k\mathbf{x}_{k-1}$ and by recurrence:
$$\mathbf{x}_k = \mathbf{M}_k\mathbf{M}_{k-1}...\mathbf{M}_1\mathbf{x}_1. \tag{2.66}$$

2. Tangent linear
   The cost function can be squared assuming that the operator $\mathbf{M}$ can be linearized, that is:
$$\mathbf{y}_k - \mathbf{H}_k\mathbf{M}_{0\to k}(x) \approx \mathbf{y}_k - \mathbf{H}_t\mathbf{M}_{0\to k}(\mathbf{x}^b) - \mathbf{H}_k\mathbf{M}_{0\to k}(\mathbf{x} - \mathbf{x}^b), \tag{2.67}$$
   where $\mathbf{M}$ is the tangent linear model, i.e. the differential of $\mathbf{M}$.

These two assumptions simplify the problem of minimizing an unconstrained quadratic function which is numerically much easier to solve. The second term of the cost function (2.64) is no more complicated than that is in $3D - Var$. The evaluation of the first term $\mathcal{J}_o$ requires N integrations of the forecasting model from the time of the analysis at each observation time k and more for the calculation of the gradient.

The evaluation of the cost function and its gradient require an integration model from $k = 0$ at the value N and the use of integration with adjoint operators performed with the transpose of the tangent linear model of temporal operators $\mathbf{M}$.[3] The minimization of the cost function and its gradient is carried out according to the following procedure:

[3]The adjoint operators have been introduced to reduce the size and the number of multiplication of matrices and to be able to calculate the cost function. Algebraically means replace a set of matrices with their transposed, hence the name of *adjoint* techniques.

1. estimate $\mathbf{x}_k = \mathbf{M}_k\mathbf{M}_{k-1}...\mathbf{M}_1 x_1$;
2. normalized discrepancies $\mathbf{d}_k = \mathbf{R}_k^{-1}(\mathbf{y}_k - \mathbf{H}_k[\mathbf{x}_k])$ that are stored
3. the contribution of the first part of cost function $\mathcal{J}_{ok} = (\mathbf{y} - \mathbf{H}_k[\mathbf{x}_k])^T \mathbf{d}_k$
4. and finally $\mathcal{J}_o = \sum_{t=0}^{N} \mathcal{J}_{ok}(\mathbf{x})$.

In order to compute $\nabla \mathcal{J}_o$ we need to factorize:

$$
\begin{aligned}
-\frac{1}{2}\nabla \mathcal{J}_o &= -\frac{1}{2}\sum_{k=0}^{N} \nabla \mathcal{J}_{ok} \\
&= \sum_{k=0}^{N} \mathbf{M}_1^T ... \mathbf{M}_k^T \mathbf{H}_k^T \mathbf{d}_k \\
&= \mathbf{H}_0^T \mathbf{d}_0 + \mathbf{M}_k^T [\mathbf{H}_1^T \mathbf{d}_1 + \mathbf{M}_1^T [\mathbf{H}_2^T \mathbf{d}_2 + ..\mathbf{H}_n^T \mathbf{d}_n]...].
\end{aligned} \tag{2.68}
$$

This equation needs to be evaluated using the following algorithm.

1. initialize the adjoint variable $\tilde{\mathbf{x}} = 0$
2. for each step $k-1$ the variable $\tilde{\mathbf{x}}_{k-1}$ is obtained adding the adjoint forcing $\mathbf{H}_k^T \mathbf{d}_k$ to $\tilde{\mathbf{x}}_k$ and performing the adjoint integration by multiplying the result for $\mathbf{M}_k^T$, i.e. $\tilde{\mathbf{x}}_{k-1} = \mathbf{M}_k^T (\tilde{\mathbf{x}}_k + \mathbf{H}_k^T \mathbf{d}_k)$
3. at the end of recurrent the adjoint value $\tilde{\mathbf{x}}_0 = -\frac{1}{2}\mathcal{J}_o(\mathbf{x})$ gives the required result.

2.3 Assimilation as an Inverse Problem

Before showing why the assimilation is an inverse problem let us introduce the concept of well and ill posed problem in the sense of Hadamard. Given and operator **A**, we wish to solve the following system of equations

$$
\mathbf{g} = \mathbf{A}\mathbf{f}. \tag{2.69}
$$

By Hadamard [9] e Hilbert e Courant [10] it is a well-posed problem when:

1. a solution exists;
2. the solution is only determined by the input parameters (forcing, boundary conditions, initial conditions);
3. depends continually on input parameters and it is stable ($\mathbf{A}^{-1}$ continuous).

When the conditions 2 and/or 3 are not satisfied the problem is ill-posed. In finite dimension, existence and uniqueness can be imposed and stability follows, however, the discrete problem of underlying ill-posed problem become ill-conditioned and the singular value of **A** decay to zero.

2.3.1 An Illustrative Example

Let us now apply the approach to a simplified model or toy model representing the sea circulation in a well defined ocean basin (Bennet [11]):

$$\frac{\partial u}{\partial t} + c\frac{\partial u}{\partial x} = F, \tag{2.70}$$

with dimension between $0 \leq x \leq L$ in the time interval $0 \leq t \leq T$ where c is a known positive constant. Let $F = F(x,t)$ be the forcing field. An initial condition is given by:

$$u(x,0) = I(x), \tag{2.71}$$

where $I(x)$ is specified. The boundary conditions are:

$$u(0,t) = B(t), \tag{2.72}$$

where $B(t)$ is defined. In order to evaluate the uniqueness of the solution we assume F, I e B have two solutions u_1 and u_2. Defining the difference $v = u_1 - u_2$ we have:

$$\frac{\partial v}{\partial t} + c\frac{\partial v}{\partial x} = 0, \tag{2.73}$$

with the boundary conditions and initial conditions respectively of $v(x,0) = 0$ and $v(0,t) = 0$. The solution can be obtained using the methods of characteristics by which Partial Differential Equations (PDEs) are reduced to Ordinary Differential Equation (ODEs). The characteristic equations are:

$$\begin{aligned} \frac{dx}{ds} &= c \\ \frac{dt}{ds} &= 1. \end{aligned} \tag{2.74}$$

The PDE that has been transformed into ODE is

$$\frac{du}{ds} = 0 \tag{2.75}$$

On the basis of the initial condition and the boundary conditions the solution is:

$$u(x,t) = 0 \tag{2.76}$$

and then $u_1(x,t) = u_2(x,t)$ showing the solution is unique.

Let us now verify the other two points of the well-posed conditions. Using the Green function one defines $G = G(x, t, \zeta, \tau)$

$$-\frac{\partial G}{\partial s} - c\frac{\partial G}{\partial x} = \delta(x - \zeta)\delta(t - \tau) \tag{2.77}$$

where δ is the Dirac's delta with $0 \leq \zeta \leq L$ and $0 \leq \tau \leq T$. The boundary conditions for $0 \leq x \leq L$ are $G(L, t, \zeta, \tau) = 0$ and for $0 \leq t \leq T$ are $G(x, T, \zeta, \tau) = 0$.

The solution is:

$$u(x,t) = \int_0^T d\tau \int_0^L d\zeta G(\zeta, \tau, x, t)F(\zeta, \tau) + \int_0^L d\zeta G(\zeta, 0, x, t)I(\zeta) + \int_0^T d\tau G(0, \tau, x, t)B(\tau) \tag{2.78}$$

that is an explicit solution for the forward model. Relation (2.78) indicates u depends on F, I, B with continuity and if they change of a $\mathcal{O}(\mathbf{e})$, u also consequently changes. Furthermore the request is that $I(0) = B(0)$ otherwise u discontinues along $x = ct$ for all ts. On the basis of such evaluation one can deduce the model is well posed.

Let us see what happens to the forward model when we introduce the information for the field $u(x, t)$ of the circulation model proposed. This information consist of imperfect observations around at an isolated point in space and time. The direct model becomes indeterminate and cannot be solved with a smooth function and therefore must be considered an ill-posed problem that must be resolved through a *best fit* weighted with all the information we hold.

Let us assume to collect a M number of measurements (observations, data, etc. ..) of u, in our basin with $0 \leq x \leq L$, during the time cruise $0 \leq t \leq T$. The data are collected in x_i, t_i with $0 \leq i \leq M$ and will be indicated by the recorded value u_i and its error as:

$$u_i = u(x_i, t_i) + e_i, \tag{2.79}$$

where $\mathbf{e}_i$ is the measurement error and $u(x_i, t_i)$ is the true value. Since the boundary conditions and the initials conditions are affected by errors the equation should be written taking into account the error f on the forcing F.

$$\frac{\partial u}{\partial t} + c\frac{\partial u}{\partial x} = F + f, \tag{2.80}$$

with

$$u(x, 0) = I(x) + i(x) \tag{2.81}$$

and

$$u(0, t) = B(t) + b(t). \tag{2.82}$$

The problem is now to obtain a unique solution for each choice of $F + f$, $I + i$ and $B + b$. This can be done by looking for the field $\hat{u} = \hat{u}(x_i, t_i)$ that minimizes errors.

One will try the minimum of the function $\mathcal{J}$ cost or penalty function where one also has introduced the error standard deviations of the "a priori" functions: W_i for the model, W_b for the boundary conditions and W_o for the observations.

$$\mathcal{J} = \mathcal{J}[u] = W_f \int_0^T dt \int_0^L f(x,t)^2 dx + W_i \int_0^L i(x)^2 dx + W_b \int_0^T b(t)^2 dt + w \sum_{m=1}^{M} e_i^2, \tag{2.83}$$

where W_f, W_i, W_b and w are the positive weights. The cost function $\mathcal{J}[u]$ is a number for each choice of the entire field u. Rewriting (2.83) and highlighting the explicit dependence of m, b, o and, e_i from F, I, B and u_i, we have:

$$\mathcal{J}(u) = W_f \int_0^T dt \int_0^L \{\frac{\partial u}{dt} + c\frac{\partial u}{\partial x} - F\}^2 dx + W_i \int_0^L \{u(x,0) - I(x)\}^2 dx +$$
$$W_b \int_0^T \{u(0,t) - B(t)\}^2 dt + w \sum_{m=1}^{M} \{u(x_i, t_i) - u_i\}^2, \tag{2.84}$$

the solution of which can be obtained using the calculation of variations as reported in Appendix, both for the solution with weak constraint and strong constraints. Since $\mathcal{J}$ is quadratic in u, it is non negative and the local extremum must be the global minimum [11].

References

1. Petersen, D.P., Middleton, D.: On representative observations. Tellus XV 4 (1963)
2. Lorenc, A.C.: The potential of the ensemble Kalman filter for NWP—a comparison with 4D-VAR. Q. J. R. Meteorol. Soc. **129**, 3183–3203 (2003)
3. Cressman, G.: An optimal objective analysis system. Mon. Weather Rev. **87**, 367–374. Part II: Data Assimilation European Centre for Medium-Range Weather Forecasts ECMWF Shinfield IFS DOCUMENTATION Cy40r1 Park, Reading, RG2 9AX, England (1959)
4. Bouttier, F., Courtier, P.: Data assimilation concepts and methods. Meteorological Training Course. ECMWF, Reading (1999)
5. Press, W.H., Flannery, B.P., Teukolsky, S.A., Vetterling, W.T.: Numerical Recipes. Cambridge University Press, Cambridge (1986)
6. Nocedal, J.: Updating quasi-Newton matrices with limited storage. Math. Comput. **35**(151), 773–782 (1980)
7. Byrd, R., Nocedal, J., Schnabel, R.: Representations of quasi-Newton matrices and their use in limited memory methods. Math. Program. **63**(4), 129–156 (1994)
8. Fisher, M., Courtier, P.: Estimating the covariance matrices of analysis and forecast error in variational data assimilation. ECMWF Tech. Memo. No. 220 (1995)
9. Hadamard, J.: An Essay on the Psychology of Invention in the Mathematical Field by Jacques Hadamard. Dover Publications, New York (1954)
10. Hilbert, D., Courant, R.: Methods of Mathematical Physics, pp. 1–2. Wiley Classical Library (1962)
11. Bennet, A.F.: Inverse Modeling of the Ocean And Atmosphere. Cambridge University Press, Cambridge (2002)

Chapter 3
Sequential Interpolation

Abstract In this chapter, the stochastic dynamical system describing the evolution of a physical system is analyzed. The Kalman filter genesis and its subsequent evolutions are outlined to give a complete overview of this algorithm. The Extended Kalman Filter, the Sigma Point Kalman Filter and the Unscented Kalman Filter are also reported. The more recent and advanced filters as the Ensemble Kalman Filters will be reported in the next chapter.

In the previous chapter, we focused our attention on the optimal estimator as well as its uncertainties, given some a priori information, the background, and the observation set, but we have not mentioned the temporal dimension of the problem. However, our interest lies in the evolution of forward model for the system state defined between times t_k and t_{k+1}. In this chapter, the stochastic dynamical system describing the evolution of a physical system is analyzed.

3.1 An Effective Introduction of a Kalman Filter

The Kalman filter is an algorithm that, given a set of measurements, arrives at the estimate of a given quantity by a recursive process. To achieve this objective, the filter processes all possible information, with no regard to their accuracy by using:

1. the knowledge of the system and its dynamic;
2. the statistical description of the noise associated with the system, the measurement error and the uncertainty of dynamic models used;
3. all possible information about the initial conditions of the variables of interest.

The concept of optimal estimate refers to the fact that the Kalman [1] filter is the best estimate that can be done, based on all information we can obtain using the three items mentioned above. The concept of recursive indicates that it is not necessary to store all information and previous measurements and reprocess whenever a new measurement has been taken. An example of how the Kalman filter works, is given in Fig. 3.1. The Kalman filter combines all available measurement data plus the a priori knowledge about the system and measurement devices to produce an estimate of the desired variables, so that the error is statistically minimized.

R. Guzzi, *Data Assimilation: Mathematical Concepts and Instructive Examples*, SpringerBriefs in Earth Sciences, DOI 10.1007/978-3-319-22410-7_3

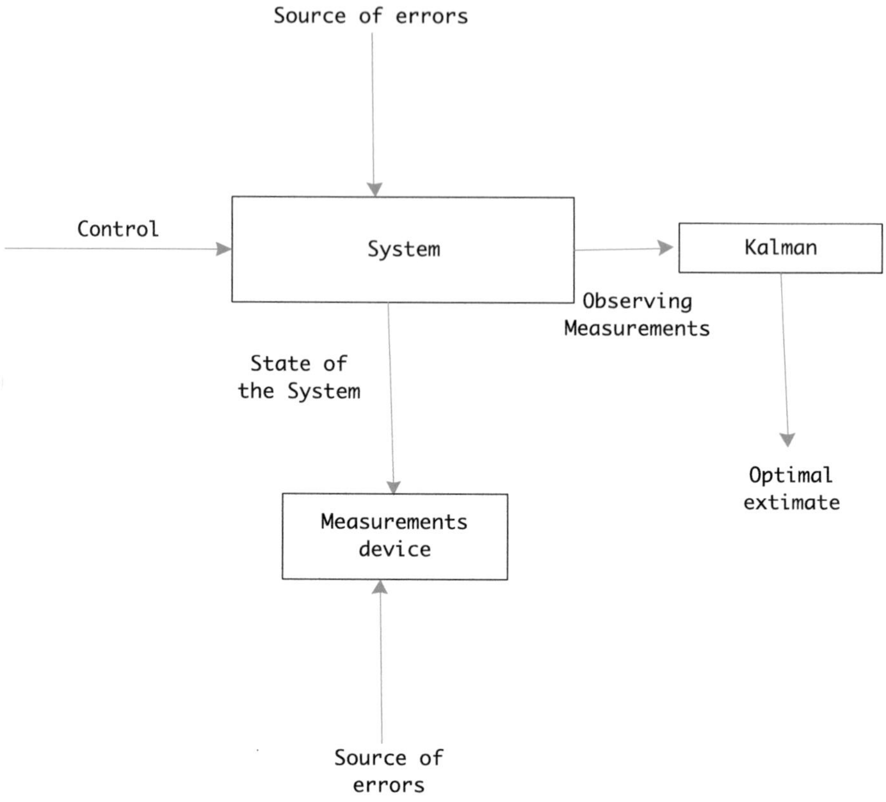

Fig. 3.1 Kalman filter process. The Kalman filter keeps track of the estimated state of the system and its variance or uncertainty. The updated estimate is obtained by a state transition model and measurements

The original derivation of the Kalman filter was not however in terms of Bayesian Rules and did not require the use of any function of probability. The only assumption was that the system of random variables should be estimated through the sequential updating of moments of first and second order (mean and covariance) and the shape of the estimator should be linear.

However, since the best approach is the one defined by the Bayesian inference, we will introduce this concept in the most effective way possible, by delaying the most complete version to the next chapters. Then, from a Bayesian point of view, the Kalman filter propagates the probability density of the given quantities conditioned by the knowledge of the actual data which comes from measurement devices. The term *conditioned* is associated with the probability density showing that its shape and its location on the axis of x depends on the values of the measurements taken. The shape of the conditioned probability density, contains the level of uncertainty, given by the variance σ, which we have given knowledge of the value of x, including the mean, the median and the mode. The Kalman filter propagates the probability density for those problems where the system can be described by a linear model in which the system error and the noise are white and Gaussian. Under these conditions the mean, the mode and the median are the same.

3.1.1 Linear System

Let us introduce the definition of a linear and nonlinear system, together with the concept of state space. Its representation is given by a mathematical model of a physical system with a data set of input, output and state variables related to differential equations of first degree. The representation of states space is a convenient and compact way to define a pattern and a test system with multiple input and output. It is built with the state variables of the system dynamic as coordinates. In engineering it is called the state space, whereas in physics it is called phase space.

The state of a system, at any time, is represented by a point in space. Starting from an initial position the point, corresponding to the state, moves in space and this movement is completely determined by the equations of state. The path of the point is called the orbit or trajectory of the system. It starts from given conditions. These trajectories are obtained from the solutions of the state equations.

A continuous-time linear system can be described by a first-order differential equations with the associated output:

$$\begin{aligned}\dot{\mathbf{x}}(t) &= \mathbf{F}\mathbf{x}(t) + \mathbf{G}\mathbf{u}(t)\\ \mathbf{y}(t) &= \mathbf{C}\mathbf{x}(t),\end{aligned} \tag{3.1}$$

where $\mathbf{x}(t)$ is a state vector; $\mathbf{F}$ is the matrix of the system; $\mathbf{G}$ is the input matrix and $\mathbf{C}$ is the output matrix. All these matrices have the appropriate sizes. The dot on $\mathbf{x}$ indicates the time derivative; $\mathbf{u}$ is the control vector and $\mathbf{y}$ is the output vector. Even though the matrices are time dependent the system is still linear.

If $\mathbf{F}$, $\mathbf{G}$, $\mathbf{C}$ are constant the solution of the Eq. (3.1) is:

$$\begin{aligned}\mathbf{x}(t) &= \exp[\mathbf{F}(t - t_0)]\mathbf{x}(t_0) + \int_{t_0}^{t} \exp[\mathbf{F}(t - \tau)]\mathbf{G}\mathbf{u}(\tau)]d\tau\\ \mathbf{y}(t) &= \mathbf{C}\mathbf{x}(t),\end{aligned} \tag{3.2}$$

where t_0 is the initial value of the system that is often zero. In case of input equal zero the relation (3.2) is:

$$\mathbf{x}(t) = \exp[\mathbf{F}(t - t_0)]\mathbf{x}(t_0). \tag{3.3}$$

If $\mathbf{x}$ is an n-dimensional vector, (3.2) is still the solution of (3.1).

The problem is now to compute $\exp[\mathbf{F}t]$. Molner and Van Loan [2] give some solutions. Among them we select the Taylor series

$$\exp[\mathbf{F}t] = \sum_{j=0}^{\infty} \frac{(\mathbf{F}t)^j}{j!}. \tag{3.4}$$

The computer generally transforms a continuous-time linear system into a discrete-time linear system. Thus, the Eq. (3.2) is transformed into discrete-time equation $t = t_k$ with the initial condition $t_0 = t_{k-1}$. Assuming that $\mathbf{F}(\tau)$, $\mathbf{G}(\tau)$ and $\mathbf{u}(\tau)$ are considered constants, in the integration interval, we obtain:

$$\mathbf{x}(t_k) = \exp[\mathbf{F}(t_k - t_{k-1})]\mathbf{x}(t_{k-1}) + \int_{t_{k-1}}^{t_k} \exp[\mathbf{F}(t_k - \tau)]\mathbf{G}\mathbf{u}(t_{k-1})d\tau. \tag{3.5}$$

Let us define now $\Delta t = t_k - t_{k-1}$ and $\alpha = \tau - t_{k-1}$; substituting into (3.5) we obtain:

$$\begin{aligned}\mathbf{x}(t_k) &= \exp[\mathbf{F}\Delta t]x(t_{k-1}) + \int_0^{\Delta t} \exp[\mathbf{F}(\Delta t - \alpha)]\mathbf{G}u(t_{k-1})d\alpha \\ &= \exp[\mathbf{F}\Delta t]x(t_{k-1}) + \exp[\mathbf{F}\Delta t]\int_0^{\Delta t} \exp[-\mathbf{F}\alpha]\mathbf{G}\mathbf{u}(t_{k-1})d\alpha. \end{aligned} \tag{3.6}$$

At this point, it is necessary to compute the integral of the exponential of the matrix. It can be simplified if $\mathbf{G}$ is invertible.

$$\begin{aligned}\int_0^{\Delta t} \exp[-\mathbf{F}\alpha]d\alpha &= \int_0^{\Delta t} \sum_{j=0}^{\infty} \frac{(-\mathbf{F})^j}{j!} d\alpha \\ &= [\mathbf{I} - \exp[-\mathbf{F}\Delta t]]\mathbf{F}^{-1}. \end{aligned} \tag{3.7}$$

Putting $\mathbf{A} = \exp[\mathbf{F}\Delta t]$ e $\mathbf{B} = \mathbf{A}[\mathbf{I} - \exp[-\mathbf{F}\Delta t]]\mathbf{F}^{-1}\mathbf{G}$ into (3.6), where Δt is the set of discretization, we can write the following dynamic discrete-time equation.

$$\mathbf{x}_k = \mathbf{A}_{k-1}\mathbf{x}_{k-1} + \mathbf{B}_{k-1}\mathbf{u}_{k-1}, \tag{3.8}$$

where we have substituted with t_k and t_{k-1} the time step previously defined with k e $k - 1$.

3.1.2 Building up the Kalman Filter

Let us assume now that the discretized dynamical system is the one relating (3.8) to which we have added $\mathbf{w}_{k-1}$, which is the white noise associated with the process. Thus we have:

$$\mathbf{x}_k = \mathbf{A}_{k-1}\mathbf{x}_{k-1} + \mathbf{B}_{k-1}\mathbf{u}_{k-1} + \mathbf{w}_{k-1}, \tag{3.9}$$

where $\mathbf{x}$, $\mathbf{u}$ and $\mathbf{w}$ are the state vector of the process at time $k-1$ with size $n \times 1$; $\mathbf{A}$ is the matrix of transition of the process state from $k-1$ to t, time stationary and with size $n \times m$. $\mathbf{B}$ is the input matrix; $\mathbf{u}_{k-1}$ is the input vector at time step $k-1$; $\mathbf{w}_{k-1}$ is the uncertainty. The index k gives the state of the system at time k.

The observation vector is:

$$\mathbf{y}_k = \mathbf{H}_k \mathbf{x}_k + \mathbf{v}_k, \tag{3.10}$$

where $\mathbf{y}_k$ defines the observation related to the measurement $\mathbf{x}$ at time k with size $m \times 1$; $\mathbf{H}_k$ is the matrix jointing the vector state to the measurement vector. For any measurements it can be changed, but usually it is taken stationary. It does not contain the noise and its size is $m \times n$; $\mathbf{v}_k$ is the measurement error and it has zero cross-correlation with the noise. Its size is $m \times 1$; $\mathbf{w}_k \sim N(0, Q_k)$ and $\mathbf{v}_k \sim N(0, R_k)$.

In summary: the matrix $\mathbf{A}$ makes the link between the current state with the previous one; $\mathbf{B}$ is the process control; $\mathbf{H}_k$ is the measurement operator. Let us now define with $\mathbf{x}^f$ the a priori estimate or in Bayesian terms.

$$\mathbf{x}_k^a = E[\mathbf{x}_k | \mathbf{y}_1, \mathbf{y}_2, \ldots \mathbf{y}_k], \tag{3.11}$$

where the a posteriori estimate is computed on the expected values of $\mathbf{x}_k$, conditioned by all measurements up to k (including k). Notice how the a priori estimate and the a posteriori, $\mathbf{x}_k^f$ and $\mathbf{x}_k^a$ are the estimates of the same quantity $\mathbf{x}_k$ respectively.

Let us define as error $\mathbf{e}^f = \mathbf{x}_k - \mathbf{x}_k^f$ and $\mathbf{e}^a = \mathbf{x}_k - \mathbf{x}_k^a$. Figure 3.2 shows it as the sequence of the a priori and a posteriori estimates and related covariances.

We have two sources of information that help us to estimate the state of the system at the time k. The first information is the dynamic equation of the system. Substituting its estimate into the relation (3.9) to the variable , we have for $\mathbf{w}_{k-1} = 0$:

$$\mathbf{x}_k^f = \mathbf{A}_{k-1} \mathbf{x}_{k-1}^a + \mathbf{B}_{k-1} \mathbf{u}_{k-1}, \tag{3.12}$$

Secondary information source is given from $\mathbf{y}_k$ own data. In order to obtain the optimal filter, we need to minimize the mean square error, provided the error of the system is Gaussian.

Let us assume models of noise are stationary in time and are evidenced by the covariance of the type:

$$E[\mathbf{w}_i \mathbf{w}_j^T] = \mathbf{Q}\delta_{ij} \tag{3.13}$$

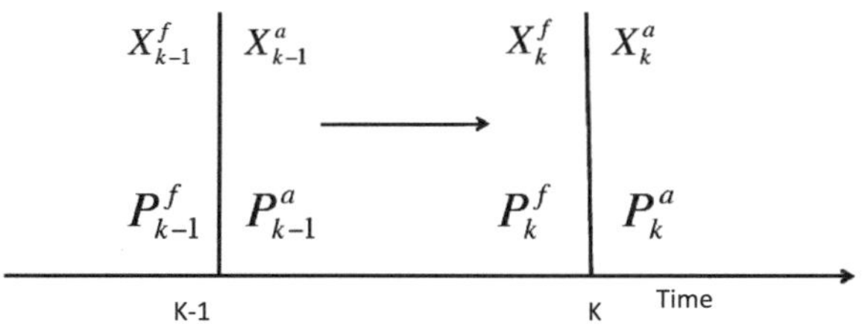

Fig. 3.2 The sequential process is based on continuous or intermittent background/forecast and analysis steps

and

$$E[\mathbf{v}_i \mathbf{v}_j^T] = \mathbf{R}\delta_{ij} \tag{3.14}$$

and

$$E[\mathbf{v}_i \mathbf{w}_j^T] = 0, \tag{3.15}$$

where δ_{ij} is the delta's Kronecker ($\delta_{ij} = 1$ if $i = j$ or $\delta_{ij} = 0$ if $i \neq j$).

The covariance matrix of the a priori error, at time t with size $n \times n$, is:

$$\mathbf{P}_k^f = E[\mathbf{e}_k^f (\mathbf{e}_k^f)^T] = E[(\mathbf{x} - \mathbf{x}_k^f)(\mathbf{x}_k - \mathbf{x}_k^f)^T] \tag{3.16}$$

and the one a posteriori

$$\mathbf{P}_k^a = E[\mathbf{e}_k^a (\mathbf{e}_k^a)^T] = E[(\mathbf{x}_k - \mathbf{x}_k^a)(\mathbf{x}_k - \mathbf{x}_k^a)^T] \tag{3.17}$$

Assuming the a priori estimate $\mathbf{x}_k^a$ is $\mathbf{x}_k^f$, (see Fig. 3.2), it is possible to write an equation, for the new estimate, that depends on a priori estimate.

$$\mathbf{x}_k^a = \mathbf{x}_k^f + \mathbf{K}_k(\mathbf{y}_k - \mathbf{H}_k \mathbf{x}_k^f), \tag{3.18}$$

where $\mathbf{K}_k$ is the Kalman gain, that we get soon, and the term $(\mathbf{y}_k - \mathbf{H}_k \mathbf{x}_k^f)$ is known as innovation. Substituting the relation (3.10) into (3.18) we get:

$$\mathbf{x}_k^a = \mathbf{x}_k^f + \mathbf{K}_k(\mathbf{H}_k \mathbf{x}_k + \mathbf{v}_k - \mathbf{H}_k \mathbf{x}_k^f), \tag{3.19}$$

that can be written as:

$$\mathbf{x}_k^a = (\mathbf{I} - \mathbf{K}_k \mathbf{H}_k)\mathbf{x}_k^f + \mathbf{K}_k \mathbf{H}_k \mathbf{x}_k + \mathbf{K}_k \mathbf{v}_k, \tag{3.20}$$

substituting it into (3.17) we get:

$$\mathbf{P}_k^a = E[(\mathbf{I} - \mathbf{K}_k \mathbf{H}_k)(\mathbf{x}_k - \mathbf{x}_k^f) - \mathbf{K}_k \mathbf{v}_k][(\mathbf{I} - \mathbf{K}_k \mathbf{H}_k)(\mathbf{x}_k - \mathbf{x}_k^f) - \mathbf{K}_k \mathbf{v}_k]^T]. \tag{3.21}$$

Since the a priori error estimate $(\mathbf{x}_k - \mathbf{x}_k^f)$ is not correlated with the measurement noise, the expectation can be written as

$$\mathbf{P}_k^a = (\mathbf{I} - \mathbf{K}_k \mathbf{H}_k)E[(\mathbf{x}_k - \mathbf{x}_k^f)(\mathbf{x}_k - \mathbf{x}_k^f)^T](\mathbf{I} - \mathbf{K}_k \mathbf{H}_k)^T + \mathbf{K}_k E[\mathbf{v}_k \mathbf{v}_k^T]\mathbf{K}_k^T. \tag{3.22}$$

Taking into account the relation (3.14) we get the update covariance equation

$$\mathbf{P}_k^a = (\mathbf{I} - \mathbf{K}_k \mathbf{H}_k)\mathbf{P}_k^f(\mathbf{I} - \mathbf{K}_k \mathbf{H}_k)^T + \mathbf{K}_k \mathbf{R} \mathbf{K}_k^T, \tag{3.23}$$

where $\mathbf{P}_k^f$ is the a priori estimate of $\mathbf{P}_k^a$.

As it is known, the diagonal of the covariance matrix contains the average quadratic error and because the trace of the matrix is the sum of the diagonal values of a matrix it will show the trace of the covariance matrix is the sum of the mean square errors. Then the mean square error can be minimized by minimizing the trace of the covariance matrix. First of all we make the derivative of $\mathbf{P}_k^a$ with respect to $\mathbf{K}_k$ and then we put it equal to zero.

Expanding the Eq. (3.23) we get:

$$\mathbf{P}_k^a = \mathbf{P}_k^f - \mathbf{K}_k\mathbf{H}_k\mathbf{P}_k^f - \mathbf{P}_k^f\mathbf{H}_k^T\mathbf{K}_k^T + \mathbf{K}_k(\mathbf{H}_k\mathbf{P}_k^f\mathbf{H}_k^T + \mathbf{R})\mathbf{K}_k^T. \tag{3.24}$$

Noting the trace of a matrix ($Trace[\cdot]$) is equal to the trace of its transpose, we can write:

$$Trace[\mathbf{P}_k^a] = Trace[\mathbf{P}_k^f] - 2Trace[\mathbf{K}_k\mathbf{H}_k\mathbf{P}_k^f] + Trace[\mathbf{K}_k(\mathbf{H}_k\mathbf{P}_k^f\mathbf{H}_k^T + \mathbf{R})\mathbf{K}_k^T]. \tag{3.25}$$

Making the derivative for $\mathbf{K}_k$ and putting it equal zero we get:

$$\frac{d\ Trace[\mathbf{P}_k^a]}{d\mathbf{K}_k} = -2(\mathbf{H}_k\mathbf{P}_k^f)^T + 2\mathbf{K}_k(\mathbf{H}_k\mathbf{P}_k^f\mathbf{H}_k^T + \mathbf{R}) = 0. \tag{3.26}$$

from which we get the Kalman gain matrix

$$\mathbf{K}_k = \mathbf{P}_k^f\mathbf{H}_k^T(\mathbf{H}_k\mathbf{P}_k^f\mathbf{H}_k^T + \mathbf{R})^{-1} \tag{3.27}$$

and the associated covariance forecast matrix

$$\mathbf{S}_k^f = (\mathbf{H}_k\mathbf{P}_k^f\mathbf{H}_k^T + \mathbf{R})^{-1}. \tag{3.28}$$

The update covariance is obtained from (3.23) in which we have substituted the Kalman gain matrix that is given by (3.27);

$$\begin{aligned}\mathbf{P}_k^a &= \mathbf{P}_k^f - \mathbf{P}_k^f\mathbf{H}_k^T(\mathbf{H}_k\mathbf{P}_k^f\mathbf{H}_k^T + \mathbf{R})^{-1}\mathbf{H}_k\mathbf{P}_k^f \\ &= (\mathbf{I} - \mathbf{K}_k\mathbf{H}_k)\mathbf{P}_k^f.\end{aligned} \tag{3.29}$$

This result gives us the update relation of the covariance matrix with optimal gain.

Equations (3.20), (3.19), (3.29) give an estimate of the variable $\mathbf{x}_k$. The projection of the next state is obtained from:

$$\mathbf{x}_{k+1}^f = \mathbf{A}_k\mathbf{x}_k^a + \mathbf{B}_k\mathbf{u}_k. \tag{3.30}$$

To complete the recursion, we need to map the error covariance matrix on the next step $k + 1$. We get it defining an expression for the a priori error:

$$\mathbf{e}_{k+1}^f = \mathbf{x}_{k+1} - \mathbf{x}_{k+1}^f = (\mathbf{A}_k\mathbf{x}_k + \mathbf{B}_k\mathbf{u}_k + \mathbf{w}_k) - (\mathbf{A}_k\mathbf{x}_k^a + \mathbf{B}_k\mathbf{u}_k) = \mathbf{A}_k\mathbf{e}_k^a + \mathbf{w}_k. \tag{3.31}$$

Extending the definition of $\mathbf{P}_k^f$ at time $k+1$ we get:

$$\mathbf{P}_{k+1}^f = E[\mathbf{e}_{k+1}^f(\mathbf{e}_{k+1}^f)^T] = E[(\mathbf{A}_k\mathbf{e}_k^a)(\mathbf{A}_k\mathbf{e}_k^a)^T] + E[\mathbf{w}_k\mathbf{w}_k^T] = \mathbf{A}_k\mathbf{P}_k^a\mathbf{A}_k^T + \mathbf{Q}, \quad (3.32)$$

that completes the recursive filter. Let us note that $\mathbf{e}_k$ and $\mathbf{w}_k$ have zero cross-correlation because the noise $\mathbf{w}_k$ accumulates between $k-1$ and k while the error $\mathbf{e}_k$ is the current error up to time k.

In summary, the Kalman filter is an algorithm that, given an initial estimate produces a gain, the so-called Kalman gain, which generates an updated estimate that generates an updated variance that is projected onto the next step.

In practice the filter operates sequentially in k; at time k we make an estimate of $\mathbf{x}_k$, $\mathbf{x}_k^a$, with a covariance matrix of the error of $\mathbf{S}_k^f$. The Eq. (3.28) defines the Kalman gain. The stochastic Eq. (3.9) is used to build the a priori estimate and its variance at the time $k+1$. This result is then combined with the measurement made at the same time, using the equation of the maximum estimate to produce an update status. The Kalman gain matrix is functionally identical to the maximum a posteriori estimate (MAP).

The algorithm can be synthetized (without the current index on **A**, **B**, **H**):

1. be:

$$\mathbf{x}_k^f = \mathbf{A}\mathbf{x}_{k-1}^a + \mathbf{B}\mathbf{u}_{k-1}; \quad (3.33)$$

2. be:

$$\mathbf{P}_k^f = \mathbf{A}\mathbf{P}_{k-1}^a\mathbf{A}^T + \mathbf{Q}; \quad (3.34)$$

3. get:

$$\mathbf{K}_k = \mathbf{P}_k^f\mathbf{H}^T(\mathbf{H}\mathbf{P}_k^f\mathbf{H}^T + \mathbf{R})^{-1}; \quad (3.35)$$

4. update:

$$\mathbf{x}_k^a = \mathbf{x}_k^f + \mathbf{K}_k(\mathbf{y}_k - \mathbf{H}\mathbf{x}_k^a); \quad (3.36)$$

5. update:

$$\mathbf{P}_k^a = (\mathbf{I} - \mathbf{K}_k\mathbf{H})\mathbf{P}_k^f. \quad (3.37)$$

3.2 More Kalman Filters

3.2.1 The Extended Kalman Filter

There are some cases in which the linearity hypothesis is not valid. For instance, when the observation operator is nonlinear as it is in the case of satellite observation, where the radiance operators are taken into account, or in the case of a forward nonlinear model as it is in atmospheric chemistry.

This means we need to adapt the Kalman filter to handle the potential nonlinearity of these operators. In these cases we can use the Extended Kalman Filter (EKF). The evolution of the state of the system is given by (3.8), which in discrete form is:

$$\mathbf{x}_k = \mathbf{f}(\mathbf{x}_{k-1}, \mathbf{u}_{k-1}, \mathbf{w}_{k-1}), \tag{3.38}$$

where $\mathbf{w}_{k-1}$ is a random perturbation of the system whose distribution has zero mean and covariance given by the matrix $\mathbf{Q}_k$.

The measurement is given by:

$$\mathbf{y}_k = \mathbf{h}(\mathbf{x}_k, \mathbf{v}_k). \tag{3.39}$$

The forecast step is like the one we have previously done for the Kalman filter.

$$\mathbf{x}_k^f = \mathbf{f}(\mathbf{x}_{k-1}^a, \mathbf{u}_{k-1}, 0). \tag{3.40}$$

Introducing the following notation for the forecast observation we have:

$$\mathbf{y}_k^f = \mathbf{h}(\mathbf{x}_k^f, 0). \tag{3.41}$$

If we argue that $\mathbf{f}(\mathbf{x}, \mathbf{u}, \mathbf{w})$ is linearly approximated, for small variations of $\mathbf{x}$ and $\mathbf{w}$, we can assume that $\mathbf{y}_k^f$ is approximately normally distributed.

Thus we can render linear $\mathbf{f}(\mathbf{x}, \mathbf{u}, \mathbf{w})$ around $(\mathbf{x}_{k-1}^a, \mathbf{u}_{k-1}, 0)$ and we obtain:

$$\mathbf{f}(\mathbf{x}, \mathbf{u}, \mathbf{w}) = \mathbf{f}(\mathbf{x}_{k-1}^a, \mathbf{u}_{k-1}, 0) + \mathbf{A}_{k-1}(\mathbf{x} - \mathbf{x}_{k-1}^a) + \mathbf{W}_{k-1}(\mathbf{w} - 0). \tag{3.42}$$

Note we assume $\mathbf{u}_k$ is known and then it has not to be linearized. $\mathbf{A}_{k-1}$ and $\mathbf{W}_{k-1}$ are partial differential matrices or Jacobian matrix and are:

$$\mathbf{A}_{i,j,k} = \frac{\partial \mathbf{f}_i(\mathbf{x}_{k-1}^a, \mathbf{u}_{k-1}, 0)}{\partial \mathbf{x}_j}. \tag{3.43}$$

$$\mathbf{W}_{i,j,k} = \frac{\partial \mathbf{f}_i(\mathbf{x}_{k-1}^a, \mathbf{u}_{k-1}, 0)}{\partial \mathbf{w}_j}. \tag{3.44}$$

With this approach we have the approximated covariance matrix for $\mathbf{x}_k^f$:

$$\mathbf{P}_k^f = \mathbf{A}_{k-1}\mathbf{P}_{k-1}^a\mathbf{A}_{k-1}^T + \mathbf{W}_{k-1}\mathbf{Q}_{k-1}\mathbf{W}_{k-1}^T. \tag{3.45}$$

Similarly we may linearize $\mathbf{h}$ placing

$$\mathbf{H}_{i,j,k} = \frac{\partial \mathbf{h}_i(\mathbf{x}_k^f, 0)}{\partial \mathbf{x}_j} \tag{3.46}$$

and

$$\mathbf{V}_{i,j,k} = \frac{\partial \mathbf{h}_i(\mathbf{x}_k^f, 0)}{\partial \mathbf{v}_j} \tag{3.47}$$

Thus,

$$\mathbf{e}_{x_k}^f = \mathbf{x}_k - \mathbf{x}_k^f \tag{3.48}$$

and

$$\mathbf{e}_{z_k}^f = \mathbf{y}_k - \mathbf{y}_k^f. \tag{3.49}$$

We do not know $\mathbf{x}_k$, but we aspect that $\mathbf{x}_k - \mathbf{x}_k^f$ is relatively small. Now we are able to linearize our function $\mathbf{f}(\cdot)$ in order to obtain an approximation for $\mathbf{e}_{x_k}^f$:

$$\mathbf{e}_{x_k}^f = \mathbf{f}(\mathbf{x}_{k-1}, \mathbf{u}_{k-1}, \mathbf{w}_{k-1}) - \mathbf{f}(\mathbf{x}_{k-1}^f, \mathbf{u}_{k-1}, 0). \tag{3.50}$$

Thus the approximated value is:

$$\mathbf{e}_{x_k}^f \approx \mathbf{A}_{k-1}(\mathbf{x}_{k-1} - \mathbf{x}_{k-1}^f) + \epsilon_k, \tag{3.51}$$

where ϵ_k is a distribution as $N(0, \mathbf{W}_{k-1}\mathbf{Q}_{k-1}\mathbf{W}_{k-1}^T)$ taking into account the effect of random variable $\mathbf{w}_{k-1}$.

Similarly

$$\mathbf{e}_k^f = \mathbf{h}(\mathbf{x}_k, \mathbf{v}_k) - \mathbf{h}(\mathbf{x}_k^f, 0), \tag{3.52}$$

can be approximately linearized by:

$$\mathbf{e}_k^f \approx \mathbf{H}_k \mathbf{e}_{x_k}^f + \eta_k, \tag{3.53}$$

where the distribution of η_k is $N(0, \mathbf{V}_k\mathbf{R}_k\mathbf{V}_k^T)$.

Ideally we need to update $\mathbf{e}_k^f$ in order to obtain $\mathbf{x}_k = \mathbf{x}_k^f + \mathbf{e}_{x_k}^f$, but , since we do not know $\mathbf{e}_{x_k}^f$ we need to estimate it by the Kalman gain factor. Placing

$$\mathbf{e}_{x_k}^a = \mathbf{K}_k(\mathbf{y}_k - \mathbf{y}_k^f) \tag{3.54}$$

and after placing $\mathbf{x}_k^a = \mathbf{x}_k^f + \mathbf{e}_{x_k}^a$ and using the same derivative process like that one used for the Kalman gain we obtain the optimal gain for EKF

$$\mathbf{K}_k = \mathbf{P}_k^f \mathbf{H}_k^T (\mathbf{H}_k \mathbf{P}_k^f \mathbf{H}_k^T + \mathbf{V}_k \mathbf{R}_k \mathbf{V}_k^T)^{-1}, \tag{3.55}$$

from which we can obtain the covariance for the update estimate of $\mathbf{x}_k^a$

$$\mathbf{P}_k^a = (\mathbf{I} - \mathbf{K}_k \mathbf{H}_k)\mathbf{P}_k^f. \tag{3.56}$$

3.2.2 Sigma Point Kalman Filter (SPKF)

The central point of the Kalman filter is the propagation of a Gaussian random variable through the dynamic system. In an Extended Kalman filter the distribution of the state of a system and the related noise density are approximated by the Gaussian random variables that are then propagated through a first order linearization of non-linear system. This can introduce large errors into true mean and in the covariance of the Gaussian random variable that can lead to a divergence in the filter compromising the operation.

To address substantially the development of the new Kalman filters, particularly the filter *Sigma Point Filter*, and to what it is connected, the *Unscented Kalman filter*, we need to take the concepts of Bayesian inference. We use such an approach at the light of the applications derived from the robotic's world where the *Sigma Point Kalman Filter* was born (Van der Merwe and Wan [3]).

The probabilistic inference consists in the problem of estimating the hidden variables (states or parameters) of a system in a consistent and optimal way, once we have the information incomplete and noisy. We consider that the hidden state of the system $\mathbf{x}_k$ with initial probability density of $p(\mathbf{x}_0)$ evolves in time (we remind that we used the k index to describe the discrete time evolution) as a first-order Markov process according to the probability density of $p(\mathbf{x}_k|\mathbf{x}_{k-1})$. Once the state variable is given, the observations $\mathbf{y}_k$ are conditionally independent and are generated according to the density of conditioned probability $p(\mathbf{y}_k|\mathbf{x}_k)$. Then the dynamic model of space is given by:

$$\mathbf{x}_k = \mathbf{f}[\mathbf{x}_{k-1}, \mathbf{u}_k, \mathbf{w}_k] \tag{3.57}$$

$$\mathbf{y}_k = \mathbf{h}[\mathbf{x}_k, \mathbf{v}_k], \tag{3.58}$$

where $\mathbf{w}_k$ is the noise-related process that drives the dynamic system through the function of non-linear transition state $\mathbf{f}$. $\mathbf{v}_k$ is the noise corrupting the measurement through the non-linear observation function of $\mathbf{h}$. The state transition density $p(\mathbf{x}_k|\mathbf{x}_{k-1})$ is fully specified by $\mathbf{f}$ and the noise distribution process is given by $p(\mathbf{w}_k)$. $\mathbf{h}$ and the noise distribution $p(\mathbf{v}_k)$ indicate the observation probability $p(\mathbf{y}_k|\mathbf{x}_k)$. We assume the external input $\mathbf{u}_k$ is known. The Dynamic Space-State Model (DSSM) along with the statistics of random noise variables, as well as the a priori distributions of the state system, define a probabilistic model of how the system temporally evolves and how we can observe the evolution of the hidden state. The problem is to know, in a recursive way, how to obtain an optimal estimate of the hidden variables of the system, when incomplete and noisy observations are taken.

From a Bayesian point of view the a posteriori filtering density

$$p(\mathbf{x}_k|\mathbf{y}_{1:k}) \tag{3.59}$$

of the state, given all observations from 1 to k $(1 : k)$,

$$\mathbf{y}_{1:k} = \{z_1, z_2 \ldots z_k\}, \tag{3.60}$$

is the complete solution to the sequential probabilistic inference and allows us to compute the optimal estimate of each state through the conditioned mean.

$$\hat{\mathbf{x}}_k = E[\mathbf{x}_k|\mathbf{y}_{1:k}] = \int \mathbf{x}_k p(\mathbf{x}_k|\mathbf{y}_{1:k}) d\mathbf{x}_k. \tag{3.61}$$

Then the problem can be reformulated: how do we compute recursively the a posteriori density when there are new observations? The answer comes from the recursive Bayesian estimation algorithm.

Using the Bayesian rule and the DSSM of the system, the a posteriori density can be expanded and factorized in the following update recursive form.

$$\begin{aligned} p(\mathbf{x}_k|\mathbf{y}_{1:k}) &= \frac{p(\mathbf{y}_{1:k}|\mathbf{x}_k)p(\mathbf{x}_k)}{p(\mathbf{y}_{1:k-1})} \\ &= \frac{p(\mathbf{y}_k|\mathbf{x}_k)p(\mathbf{x}_k|\mathbf{y}_{1:k-1})}{p(\mathbf{y}_k|\mathbf{y}_{1:k-1})}. \end{aligned} \tag{3.62}$$

Let us see how the relation (3.62) is constructed: the a posteriori state at time $k-1$, $p(\mathbf{x}_{k-1}|\mathbf{y}_{1:k-1})$ is firstly forward projected in order to compute the a priori state at time k, using the model of probabilistic process

$$p(\mathbf{x}_k|\mathbf{y}_{1:k-1}) = \int p(\mathbf{x}_k|\mathbf{x}_{1:k-1}) p(\mathbf{x}_{k-1}|\mathbf{y}_{1:k-1}) d\mathbf{x}_{k-1}. \tag{3.63}$$

Then the more recent measurement of noise is incorporated, using the observed probability function, to generate a posteriori update state.

$$p(\mathbf{x}_k|\mathbf{y}_{1:k}) = Cp(\mathbf{y}_k|\mathbf{x}_k)p(\mathbf{x}_k|\mathbf{y}_{1:k-1}). \tag{3.64}$$

The normalization factor is given by:

$$C = (p(\mathbf{y}_k|\mathbf{x}_k)p(\mathbf{x}_k|\mathbf{y}_{1:k-1}))^{-1}. \tag{3.65}$$

The a priori transition state is:

$$p(\mathbf{x}_k|\mathbf{x}_{k-1}) = \int \delta(\mathbf{x}_k - \mathbf{f}[\mathbf{x}_{k-1}, \mathbf{u}_k, \mathbf{w}_k]) p(\mathbf{w}_k) d\mathbf{w}_k \tag{3.66}$$

and the probability density of the observation is given by:

$$p(\mathbf{y}_k|\mathbf{x}_k) = \int \delta(\mathbf{y}_k - \mathbf{h}[\mathbf{x}_k, \mathbf{v}_k]) p(\mathbf{v}_k) d\mathbf{v}_k, \tag{3.67}$$

where δ is the Dirac's delta. These multi-dimensional integrals can only be treated in the case of a Gaussian linear system.

The methodology proposed to solve the problem of probabilistic inference, the Bayesian optimal recursive solution, requires the propagation of probability density function of the a posteriori state. This solution is general enough to deal with all forms of a posteriori density, including the multimodality, asymmetries and discontinuity. However since the solution does not place any restrictions on the form of a posteriori density, in general it cannot be described by a finite number of parameters. Then each estimator must be approximated as a function of the form of a posteriori density and of the form of recursive Bayesian structure as has been defined previously.

A common mistake is to think the Kalman filter requires that the space in which it operates is linear and that the probability density is Gaussian. The Kalman filter does not require these conditions, but only the following assumptions.

1. Estimates of minimum variance of random variables, and then the distribution of subsequent state variables can be computed, only by computing the first and second momentum (i.e. the mean and the variance) by propagating them recursively and updating.
2. The estimate (the updated measurement) is a linear function of a priori knowledge of the system, sinthetizing $p(\mathbf{x}_k|\mathbf{y}_{1:k-1})$, and new information $p(\mathbf{y}_k|\mathbf{x}_k)$. In other words, we assume that the Eq. (3.67) of the optimal Bayesian recursion can be approximated by a linear function.
3. The accurate predictions of the state variable (using the process model) and the observations (using the forecast model) can be calculated to approximate the first and second momentum $p(\mathbf{x}_k|\mathbf{y}_{1:k-1})$ and $p(\mathbf{y}_k|\mathbf{x}_k)$.

Because the first assumption is consistent, it is necessary that the estimate of the mean and of the covariance of the a posteriori state density, $\hat{\mathbf{x}}_k$ and $\mathbf{P}_{\mathbf{x}_k}$, satisfy the following inequality:

$$Trace[\mathbf{P}_{\mathbf{x}_k} - E[(\mathbf{x} - \hat{\mathbf{x}}_k)(\mathbf{x} - \hat{\mathbf{x}}_k)^T]] \geq 0, \tag{3.68}$$

where $\mathbf{x} - \hat{\mathbf{x}}_k$ is called the error estimator .

The assumption 2 assumes that the update of the measure is linear and implying that the approximate Gaussian estimator is the best linear estimator (as defined by the minimum variance criterion in assumption 1). This in turn implies that we must calculate the forecast of assumption 3 in a optimally way.

Based on these assumptions we can derive the recursive form of Kalman filter through the conditional mean of the random state variable $\hat{\mathbf{x}}_k = E[\mathbf{x}_k|\mathbf{y}_{1:k}]$ and of its covariance $\mathbf{P}_{\mathbf{x}_k}$.

In the case of a recursive linear form, although we accept that the model is not linear, we have:

$$\hat{\mathbf{x}}_k = (forecast\ of\ \mathbf{x}_k) - \mathbf{K}_k(\mathbf{y}_k - (forecast\ of\ \mathbf{y}_k) \tag{3.69}$$
$$= \hat{\mathbf{x}}_k^- + \mathbf{K}_k(\mathbf{y}_k - \hat{\mathbf{z}}_k^-) \tag{3.70}$$
$$\mathbf{P}_{\mathbf{x}_k} = \mathbf{P}_{\mathbf{x}_k}^- - \mathbf{K}_k\mathbf{P}_{\tilde{\mathbf{z}}_k}\mathbf{K}_k^T, \tag{3.71}$$

where the recursive terms are:

$$\hat{\mathbf{x}}_k^- = E[\mathbf{f}(\mathbf{x}_{k-1}, \mathbf{w}_{k-1}, \mathbf{u}_k)] \tag{3.72}$$
$$\hat{\mathbf{z}}_k^- = E[\mathbf{h}(\mathbf{x}_k^-, \mathbf{v}_k)] \tag{3.73}$$
$$\mathbf{K}_k = E[(\mathbf{x}_k - \hat{\mathbf{x}}_k^-)(\mathbf{y}_k - \hat{\mathbf{z}}_k^-)^T E[(\mathbf{y}_k - \hat{\mathbf{z}}_k^-)(\mathbf{y}_k - \hat{\mathbf{z}}_k^-)^T] \tag{3.74}$$
$$= \mathbf{P}_{\mathbf{x}_k\tilde{\mathbf{z}}_k}\mathbf{P}_{\tilde{\mathbf{z}}_k}^{-1}, \tag{3.75}$$

with $\hat{\mathbf{x}}_k^-$ as the optimal forecast (a priori mean at the time k) of $\mathbf{x}_k$ and corresponds to the expectation (taken on the a posteriori distribution of the state variable at the time $k-1$) of a nonlinear function of the random variables $\mathbf{x}_{k-1}$ e $\mathbf{w}_{k-1}$. Similarly for the optimal forecast $\hat{\mathbf{z}}_k^-$, except that the expectation is taken on the a priori distribution of time state variable k. The term $\mathbf{K}_k$ is expressed as a function of the expectation of a cross-correlazione matrix (covariance matrix) of the state variable forecast error.

Since the problem is to carefully calculate the expected mean and the covariance of a random variable, let us see what uncertainties arise when estimates of the future state of a system or measurements are performed. If $\mathbf{x}$ is a random variable with mean $\bar{\mathbf{x}}$ and covariance $\mathbf{P}_{xx}$ we can always find a second random variable $\mathbf{y}$ that has a non-linear functional relationship with $\mathbf{x}$ of the type

$$\mathbf{y} = \mathbf{f}[\mathbf{x}]. \tag{3.76}$$

Now we want to calculate the statistics of $\mathbf{y}$ i.e. the average of $\bar{\mathbf{y}}$ and covariance $\mathbf{P}_{yy}$. We need to determine the distribution density function transformed and evaluate the statistics from this distribution. In the case of a linear function there exists an exact solution, but in the case of a nonlinear function we must find an approximate solution that is statistically significant. Ideally it should be efficient and unbiased. Because the statistic transformed is consistent it is necessary that the following inequality holds:

$$\mathbf{P}_{yy} - E[\{\mathbf{y} - \bar{\mathbf{y}}\}\{\mathbf{y} - \bar{\mathbf{y}}\}^T] \geq 0. \tag{3.77}$$

This condition is extremely important for the validity of the method of processing. If the statistic is inconsistent the value of $\mathbf{P}_{yy}$ is underestimated and therefore the Kalman filter assigns a weight too high to the information and underestimates the covariance. Then the filter tends to diverge. It is therefore appropriate that the transformation is efficient, that is that the left side of the Eq. (3.77) is minimized and that the estimate is unbiased or $\bar{\mathbf{y}} \approx E[\mathbf{y}]$.

We develop a consistent, efficientand unbiased transformation by developing a Taylor series around $\bar{\mathbf{x}}$ with the Eq. (3.76).

$$\mathbf{f}[\mathbf{x}] = \mathbf{f}[\bar{\mathbf{x}} + \delta\mathbf{x}] = \mathbf{f}[\bar{\mathbf{x}}] + \frac{\partial \mathbf{f}}{\partial \bar{\mathbf{x}}}\delta\mathbf{x} + \dots \frac{1}{n!}\frac{\partial^n \mathbf{f}}{\partial \bar{\mathbf{x}}^n}\delta\mathbf{x}^n, \tag{3.78}$$

where $\delta\mathbf{x}$ is a Gaussian variable with zero mean, with covariance $\mathbf{P}_{xx}$ and the $\frac{\partial^n \mathbf{f}}{\partial \bar{\mathbf{x}}^n}\delta\mathbf{x}^n$ is the nth-order of the multidimensional Taylor series. Taking the expectation of the transformed mean we have:

$$\bar{\mathbf{y}} = \mathbf{f}[\bar{\mathbf{x}}] + \frac{1}{2}\frac{\partial^2 \mathbf{f}}{\partial \bar{\mathbf{x}}^2}\mathbf{P}_{xx} + \frac{1}{2}\frac{\partial^4 \mathbf{f}}{\partial \bar{\mathbf{x}}^4}E[\mathbf{x}^4] + \dots. \tag{3.79}$$

and the related covariance is:

$$\mathbf{P}_{yy} = \frac{\partial \mathbf{f}}{\partial \bar{\mathbf{x}}}\mathbf{P}_{xx}\left(\frac{\partial \mathbf{f}}{\partial \bar{\mathbf{x}}}\right)^T + \frac{1}{2 \times 4!}\frac{\partial^2 \mathbf{f}}{\partial \bar{\mathbf{x}}^2}(E[\mathbf{x}^4] - E[\mathbf{x}^2\mathbf{P}_{yy}] - E[\mathbf{P}_{yy}\mathbf{x}^2] + \mathbf{P}_{yy}^2)\left(\frac{\partial^2 \mathbf{f}}{\partial \bar{\mathbf{x}}^2}\right)^T + \frac{1}{3!}\frac{\partial^3 \mathbf{f}}{\partial \bar{\mathbf{x}}^3}E[\mathbf{x}^4]\left(\frac{\partial \mathbf{f}}{\partial \bar{\mathbf{x}}}\right)^T + \dots. \tag{3.80}$$

In other words, the n order term in the series for $\bar{\mathbf{x}}$ is a function of the nth-momentum of $\mathbf{x}$ multiplied by the nth-derivative of $\mathbf{f}[\cdot]$ evaluated at $\mathbf{x} = \bar{\mathbf{x}}$. If the moments and the derivatives can be properly evaluated until the nth-order, the mean is correct up to that order. Similarly it applies to covariance, although the structure of each term is more complicated. Because each term in the series is scaled down to the smallest terms, the lowest order in the series has a bigger impact, so that our forecast procedure should be concentrated on evaluating the terms of lower order. The linearization process assumes that the second order terms and those of largest order in $\delta\mathbf{x}$ are neglected. Therefore under this assumption

$$\bar{\mathbf{y}} = \mathbf{f}[\bar{\mathbf{x}}] \tag{3.81}$$

$$\mathbf{P}_{yy} = \frac{\partial \mathbf{f}}{\partial \bar{\mathbf{x}}}\mathbf{P}_{xx}\left(\frac{\partial \mathbf{f}}{\partial \bar{\mathbf{x}}}\right)^T. \tag{3.82}$$

By comparing these estimates with quadratic equations (3.79) and (3.80) we see that, for the mean, the first order approximations are only valid when orders higher than the second are insignificant while, for the covariance matrix, those that are higher than the fourth order.

The Kalman filter calculates the optimal conditions through relations (3.75), while the Extended Kalman filter linearizes the system around the current state using a first-order-truncation of the multidimensional Taylor series. Broadly it approximates the optimal conditions as:

$$\hat{\mathbf{x}}_k^- = E[\mathbf{f}(\hat{\mathbf{x}}_{k-1}, \bar{\mathbf{w}}, \mathbf{u}_k)] \tag{3.83}$$

$$\hat{\mathbf{z}}_k^- = E[\mathbf{h}(\hat{\mathbf{x}}_k^-, \bar{\mathbf{v}})] \tag{3.84}$$

$$\mathbf{K}_k = \mathbf{P}^{lin}_{\mathbf{x}_k\tilde{\mathbf{z}}_k}(\mathbf{P}^{lin}_{\tilde{\mathbf{z}}_k})^{-1}, \tag{3.85}$$

where estimates are approximate without calculating any expectation, just as an average value. The averages of noises $\bar{\mathbf{w}}$ and $\bar{\mathbf{v}}$ are usually assumed equal zero. Furthermore covariances $\mathbf{P}^{lin}_{\mathbf{x}_k\tilde{\mathbf{z}}_k}$ e $\mathbf{P}^{lin}_{\tilde{\mathbf{z}}_k}$ are determined by linearizing the system model around the current estimate of the state variable and then approximatively determining analytically the a posteriori covariance matrices. This approach not only produces errors to the Extended Kalman filter, but also makes the filter may diverge.

The Kalman filter based on *Sigma Points*, increases the accuracy, robustness and efficiency of approximate inference Gaussian algorithms applied to a non-linear system. Its different approach is to calculate the a posteriori statistics of the first and second order of a random variable which is subject to a nonlinear transformation. The distribution of the state variable is still represented by a Gaussian random variable, but it is specified using a minimum set of sample points weighed in a deterministic way. These points, which are called sigma points, completely capture the actual mean covariance of a priori random variable. When they are then propagated through a non-linear system, they capture the a posteriori mean covariance up to the second order. The basic approach of the transformation of data using the method of *sigma* points is defined as follows:

1. A set of samples weighed (*Sigma Points*) is deterministically calculated using the mean and the decomposition of the covariance matrix of the a priori random variable. The set of sigma points to be used shall be such as to capture fully the first and second moment of the a priori random variable. If we want to capture higher moments we have to increase the sigma points.
2. Sigma points are propagated through the non-linear functions using only a functional assessment: i.e. we do not use analytic derivatives to generate the sigma points retrospectively.
3. The a posteriori statistics is calculated using the Sigma points that are propagated and of the weights that take the form of a sample of a weighted average and of the variance calculation.

In order to translate these steps in an algorithm we take a set of points *sigma*, $\chi_i = \{i = 0, 1, 2...n : x^{(i)}, W^{(i)}\}$ consisting in $n + 1$ vectors with their associated weights W_i. The weights, that can be either positive or negative, must obey to the normalization condition:

$$\sum_{i=0}^{n} W^{(i)} = 1. \tag{3.86}$$

Once we have these points and their weights, it is possible to calculate the average value $\bar{\mathbf{y}}$ and the covariance $\mathbf{P}_{yy}$, through the following process:

1. each point is transformed by a function into a set of *Sigma Points*

$$\boldsymbol{\Upsilon}^{(i)} = \mathbf{g}[\mathbf{x}^{(i)}]. \tag{3.87}$$

2. The mean is given by the weighted mean of the transformed points

$$\bar{\Upsilon} = \sum_{i=0}^{n} W^{(i)} \Upsilon^{(i)}. \tag{3.88}$$

3. The covariance is weighed using the outer product of transformed points treating Υ

$$\mathbf{P}_{\Upsilon} = \sum_{i=0}^{n} W^{(i)} \{\Upsilon^{(i)} - \bar{\Upsilon}\}\{\Upsilon^{(i)} - \bar{\Upsilon}\}^T. \tag{3.89}$$

A set of points that satisfy the above mentioned conditions is a symmetric set of $2N_x$ points that lies on the boundary of $\sqrt{N_x}$-th covariance

$$\begin{aligned}
\chi^{(0)} &= \bar{\mathbf{x}} \\
W^{(0)} &= W^{(0)} \\
\chi^{(i)} &= \bar{\mathbf{x}} + \left(\sqrt{\frac{N_x}{1 - W^{(0)}} \mathbf{P}_{xx}}\right)_i \\
W^{(i)} &= \frac{1 - W^{(0)}}{2N_x} \\
\chi^{(i+n_x)} &= \bar{\mathbf{x}} - \left(\sqrt{\frac{N_x}{1 - W^{(0)}} \mathbf{P}_{xx}}\right)_i \\
W^{(i+N_x)} &= \frac{1 - W^{(0)}}{2N_x},
\end{aligned} \tag{3.90}$$

where $(\sqrt{N_x \mathbf{P}_{xx}})_i$ is the ith column or row of the square root of the matrix $N_x \mathbf{P}_{xx}$ which is the covariance matrix multiplied by the number of dimensions. $W^{(i)}$ is the weight associated with the ith point. By convention $W^{(0)}$ is the weight of the mean point which is indexed as zero point. Note that the weights are not necessarily positive because they depend on the approach used for the sigma points.

Because SPKF does not use the Jacobian of the system of equations, it becomes particularly attractive for the “black box" systems or in expressions of the dynamic system in which they can be linearized.

Because the mean and the covariance of $\mathbf{x}$ are captured accurately up to second order, the calculated value of the mean of $\mathbf{y}$ are correct up to the second order. This indicates that the method is more accurate than the EKF with the additional benefit that the distribution has been approximated rather than the $\mathbf{f}[\cdot]$ since the series expansion has not been truncated in some order. In this way the algorithm is able to incorporate information from the highest orders giving greater accuracy.

The *Sigma Points* capture the same mean and covariance with no regard to the choice of the square root of the matrix used. We can for example use matrix decomposition methods more stable and efficient as the decomposition of Cholesky.

The mean and covariance have been calculated using the standard operations on vectors and this means that the algorithm is adaptable to any choice that has been made on the process, therefore compared to the EKF it is not necessary to evaluate the Jacobian matrix.

3.2.3 Unscented Kalman Filter (UKF)

The EKF has two problems: linearization can produce highly unstable filters if the assumption of linearity is locally violated; the derivation of the Jacobian matrix is not trivial for many applications and often leads to difficulties. In addition, because the linearization of the errors in EKF introduces an error which is about 1.5 times the standard deviation of the measurement interval, the transformation is inconsistent. In practice, this inconsistency may be resolved by introducing an additional noise that stabilizes the transformation, but induces a growth of the size of covariance transformed. This is a possible reason why EKF are difficult to adjust. In fact, we must introduce a sufficient noise to perform the linearization. However the noise stabilizing is an undesirable solution since the estimate remains biased and there is no guarantee that transformed estimate is consistent.

To overcome these drawbacks the UKF transformation is introduced. This is a method for calculating the statistic of a random variable which is subject to a nonlinear transformation. It is based on the assertion that it is easier to approximate a Gaussian distribution than approximate an arbitrary nonlinear function.

A UKF transformation is based on two fundamental points:

1. Is easier to make a non-linear transformation of a single point rather than an entire probability density function.
2. It is not too hard to find a set of individual points in the state phase whose probability density function sample approximates the actual probability density function of a state vector.

The UKF is closely linked to the transformation of the *Sigma*. The formula for the *Unscented* Kalman filter is:

1. Take the nonlinear system at n-discrete time states is given by:

$$\begin{aligned} \mathbf{x}_k &= f(\mathbf{x}_{k-1}, \mathbf{u}_{k-1}, \mathbf{w}_{k-1}) \\ \mathbf{y}_k &= \mathbf{h}(x_k, \mathbf{v}_k) \\ \mathbf{w}_k &\sim (0, \mathbf{Q}_k) \\ \mathbf{v}_k &\sim (0, \mathbf{R}_k). \end{aligned} \tag{3.91}$$

2. The UKF is initialized as:

$$\begin{aligned} \mathbf{x}_0^a &= E[x_0] \\ \mathbf{P}_0^a &= E[(\mathbf{x}_0 - \mathbf{x}_0^a)(\mathbf{x}_0 - \mathbf{x}_0^a)^T]. \end{aligned} \tag{3.92}$$

The following equations are used to make the upgrade timely in order to propagate the estimated state and covariance from a temporal measurement to another.

1. In order to propagate from the time $(k-1)$ to k, first select the sigma $\mathbf{x}_{k-1}^{(i)}$ as specified previously in the transformation of SPKF with an appropriate exchange because the current best hypothesis for the mean and the covariance of $\mathbf{x}_k$ are $\mathbf{x}_{k-1}^{a}$ and $\mathbf{P}_{k-1}^{a}$:

$$\begin{aligned} \mathbf{x}_{k-1}^{(i)} &= \mathbf{x}_{k-1}^{a} + \tilde{\mathbf{x}}^{(i)} \ i = 1, ...2n \\ \tilde{\mathbf{x}}^{(i)} &= (\sqrt{\mathbf{x}\mathbf{P}_{k-1}^{a}})_i^T \ i = 1, ...n \\ \tilde{\mathbf{x}}^{(n+i)} &= -(\sqrt{\mathbf{x}\mathbf{P}_{k-1}^{a}})_i^T \ i = 1, ...n. \end{aligned} \tag{3.93}$$

2. Use the equation of known nonlinear system $\mathbf{f}(\cdot)$ to transform the sigma points in vectors $x_k^{(i)}$ taking into account that the transformation is $\mathbf{f}(\cdot)$ rather than $\mathbf{h}(\cdot)$ data from SPKF and therefore we have to make the appropriate changes.

$$\mathbf{x}_k^{(i)} = \mathbf{f}(\mathbf{x}_{k-1}^{(i)}, \mathbf{u}_{k-1}, \mathbf{w}_{k-1}). \tag{3.94}$$

3. Combine vectors $\mathbf{x}_k^{(i)}$ to get the a priori state estimate at the time k

$$\mathbf{x}_k^{f} = \frac{1}{2n}\sum_{i=1}^{2n} \mathbf{x}_k^{(i)}. \tag{3.95}$$

4. Estimate the a priori covariance of the error adding $\mathbf{Q}_{k-1}$to account for the noise.

$$\mathbf{P}_k^{f} = \frac{1}{2n}\sum_{i=1}^{2n} (\mathbf{x}_k^{(i)} - \mathbf{x}_k^{f})(\mathbf{x}_k^{(i)} - \mathbf{x}_k^{f})^T + \mathbf{Q}_{k-1}. \tag{3.96}$$

Once we have done the upgrade, we implement the update equation of measurement

1. We choose the sigma points as specified by the unscented transform with an appropriate change because the current best hypothesis for the mean and the covariance of $\mathbf{x}_k$ are $\mathbf{x}_k^{f}$ e $\mathbf{P}_k^{f}$:

$$\begin{aligned} \mathbf{x}_k^{(i)} &= \mathbf{x}_k^{f} + \tilde{\mathbf{x}}^{(i)} \ i = 1, ...2n \\ \tilde{\mathbf{x}}^{(i)} &= (\sqrt{\mathbf{x}\mathbf{P}_k^{f}})_i^T \ i = 1, ...n \\ \tilde{\mathbf{x}}^{(n+i)} &= -(\sqrt{\mathbf{x}\mathbf{P}_k^{f}})_i^T \ i = 1, ...n. \end{aligned} \tag{3.97}$$

This passage can be neglected re-using the sigma points obtained from the time update. Let us use the equation of nonlinear system known $\mathbf{h}(\cdot)$ to transform the sigma vectors $\mathbf{y}_k^{(i)}$ according to the relation:

$$\mathbf{y}_k^{(i)} = \mathbf{h}(\mathbf{x}_k^{(i)}, \mathbf{u}_k, \mathbf{v}_k). \tag{3.98}$$

2. Combine vectors $\mathbf{y}_k^{(i)}$ to obtain the estimate of the a priori state at time k

$$\mathbf{y}_k^f = \frac{1}{2n} \sum_{i=1}^{2n} \mathbf{y}_k^{(i)}. \tag{3.99}$$

3. Estimate the covariance of the a priori error adding $\mathbf{R}_k$ to take account of the noise on the measurement

$$\mathbf{P}_z = \frac{1}{2n} \sum_{i=1}^{2n} (\mathbf{y}_k^{(i)} - \mathbf{y}_k^f)(\mathbf{y}_k^{(i)} - \mathbf{y}_k^f)^T + \mathbf{R}_k. \tag{3.100}$$

4. Estimate the cross-variance between $\mathbf{x}_k^f$ and $\mathbf{y}_k$

$$\mathbf{P}_{xz} = \frac{1}{2n} \sum_{i=1}^{2n} (\mathbf{x}_k^{(i)} - \mathbf{x}_k^f)(\mathbf{y}_k^{(i)} - \mathbf{y}_k)^T. \tag{3.101}$$

5. Finally to update the measurement of the estimate of the state we use the Kalman filter equation

$$\begin{aligned}
\mathbf{K}_k &= \mathbf{P}_{xz}\mathbf{P}_z^{-1} \\
\mathbf{x}_k^a &= \mathbf{x}_k^f + \mathbf{K}_k(\mathbf{y}_k - \mathbf{y}_k^f) \\
\mathbf{P}_k^a &= \mathbf{P}_k^f - \mathbf{K}_k\mathbf{P}_z\mathbf{K}_k^T.
\end{aligned} \tag{3.102}$$

We have assumed that the equations of the process and those of measurement are linear if compared to the noise, even though this is not generally true. In this case we must treat the state vector in a different ways, using what Julier and Uhlmann [4] call the augmented state (+).

$$\mathbf{x}_k^+ = \begin{bmatrix} \mathbf{x}_k \\ \mathbf{w}_k \\ \mathbf{v}_k. \end{bmatrix}$$

Thus the UKF is initialized as

$$\mathbf{x}_0^{+a} = \begin{bmatrix} E(x_0) \\ 0 \\ 0 \end{bmatrix}$$

$$\mathbf{P}_0^{+a} = \begin{bmatrix} E[(\mathbf{x}_0 - \mathbf{x}_0^{+a})(\mathbf{x}_0 - \mathbf{x}_0^{+a})^T] & 0 & 0 \\ 0 & Q_0 & 0 \\ 0 & 0 & R_0 \end{bmatrix}$$

In this case we need to estimate the mean as increased mean and the covariance as increased covariance as we have done by the algorithms previously defined but from which we have removed $\mathbf{Q}_{k-1}$ e $\mathbf{R}_k$.

References

1. Kalman, R.E., Bucy, R.S.: New results in linear filtering and prediction theory. Trans. Am. Soc. Mech. Eng. J. Basic Eng. Ser. D **83**, 95–108 (1961)
2. Molner, C., VanLoan, C.: Nineteen Dubious Ways to Compute the Exponential of a Matrix. Twenty-Five Years, SIAM Rev. Soc. Ind. Appl. Math. **45**, 349 (2003)
3. van der Merwe R., Wan, E.A., Julier, S.J.: Sigma-Point kalman filters for nonlinear estimation and sensor-fusion: applications to integrated navigation. In: Proceedings of the AIAA Guidance, Navigation, and Control Conference, Providence, RI (2004)
4. Julier, S., Uhlmann, J.: Unscented filtering and nonlinear estimation. In: Proceeding of IEEE **92**(3), 401–422 (2004)

Chapter 4
Advanced Data Assimilation Methods

Abstract The computations of the model and its error require an accurate selection of the methods to be used. When the algorithm adopted to specific non-linear systems diverges or the approximation order to better handle the non linearity of the model fails, we need to improve the performance of the method. This chapter deals as overcome these problems by the recursive Bayesian estimation and the most advanced filters such as the stochastic ensemble Kalman filters.

As we have seen in the previous chapters the accurate numerical prediction requires accurate initial conditions. However the computations of the model and its error require an accurate selection of the methods to be used. For example the Extended Kalman Filter applied to specific non-linear systems diverges because it is not able to gather enough information on the system state's trajectory and because the error propagation is approximated by the tangent linear model in analysis steps. The problem can partially overcome by using higher order approximation in order to better handle the non linearity of the model as, for instance happens in the Sigma Point Kalman filter (SPKF) or derived from. However, beyond the tangent linear, there is also a requirement of a Hessian which in terms of computer time is expensive.

Then in order to improve the performance of the method, with specific nonlinear systems, we need to take into account all statistical moments of the state distribution due to their impact in propagating within the system.

If we want to work with all statistics moments we must not limit ourselves to the expectations and the covariances of distributions of variables. There the problem needs to be reformulated in terms of probability of density function or *pdf* of a state system.

This approach can be addressed using Bayesian statistics and stochastic filtering. Before demonstrating the Kalman filter can be derived in the framework of the Bayesian statistics, we must introduce the recursive Bayesian estimation.

4.1 Recursive Bayesian Estimation

The recursive sequential Bayesian filtering is based on recursive Bayesian estimation. The filtering problem is a conditional a posteriori density estimation problem that, in the Hadamard sense, is a stochastically ill posed problem.

R. Guzzi, *Data Assimilation: Mathematical Concepts and Instructive Examples*,
SpringerBriefs in Earth Sciences, DOI 10.1007/978-3-319-22410-7_4

The stochastic filter can be described as giving the initial density $p(\mathbf{x}_0)$, the transition density $p(\mathbf{x}_k|\mathbf{x}_{k-1})$, and the likelihood $p(\mathbf{y}_k|\mathbf{x}_k)$. The objective of the filtering is to estimate the optimal current state at time k given the observations up to time k, which is how to arrive at estimating the a posteriori density $p(\mathbf{x}_k|\mathbf{y}_{0:k})$ or $p(\mathbf{x}_{0:k}|\mathbf{y}_{0:k})$.

Although the posterior density provides a complete solution of the filtering problem, the problem is still complex since the density is a function rather than a finite-dimensional point estimate. Because several physical systems do not have finite dimension, the infinite-dimensional system can only be approximately modeled trough a finite-dimensional filter, i.e. the filter can only be suboptimal.

In order to derive the recursive Bayesian filter, two assumptions are produced:

1. The states follow a first-order Markov process $p(\mathbf{x}_k|\mathbf{x}_{0:k-1}) = p(\mathbf{x}_k|\mathbf{x}_{k-1})$;
2. the observations are independent of the given states. For simplicity, we denote with $\mathcal{Y}_l = \mathbf{y}_{0:l} := \{\mathbf{y}_0, \ldots, \mathbf{y}_l\}$ a set of observations available up to some time t_l, while $\mathcal{Y}_k = \mathbf{y}_{0:k} = \mathbf{y}_{0:k} := \{\mathbf{y}_0, \ldots, \mathbf{y}_k\}$ is the solution of filtering problem for $t_k, k = 1, \ldots 3$

If the conditional *pdf* of $\mathbf{x}_k$ is denoted by $p(\mathbf{x}_k|\mathcal{Y}_k)$, from Bayesian rule we have

$$\begin{aligned} p(\mathbf{x}_k|\mathcal{Y}_k) &= \frac{p(\mathcal{Y}_k|\mathbf{x}_k)p(\mathbf{x}_k)}{p(\mathcal{Y}_k)} \\ &= \frac{p(\mathbf{y}_k, \mathcal{Y}_{k-1}|\mathbf{x}_k)p(\mathbf{x}_k)}{p(\mathbf{y}_k, \mathcal{Y}_{k-1})} \\ &= \frac{p(\mathbf{y}_k|\mathcal{Y}_{k-1}, \mathbf{x}_k)p(\mathcal{Y}_{k-1}|\mathbf{x}_k)p(\mathbf{x}_k)}{p(\mathbf{y}_k|\mathcal{Y}_{k-1})p(\mathcal{Y}_{k-1})} \\ &= \frac{p(\mathbf{y}_k|\mathcal{Y}_{k-1}, \mathbf{x}_k)p(\mathbf{x}_k|\mathcal{Y}_{k-1})p(\mathcal{Y}_{k-1})p(\mathbf{x}_k)}{p(\mathbf{y}_k|\mathcal{Y}_{k-1})p(\mathcal{Y}_{k-1})p(\mathbf{x}_k)} \\ &= \frac{p(\mathbf{y}_k|\mathbf{x}_k)p(\mathbf{x}_k|\mathcal{Y}_{k-1})}{p(\mathbf{y}_k|\mathcal{Y}_{k-1})}, \end{aligned} \tag{4.1}$$

where the posterior density $p(\mathbf{x}_k|\mathcal{Y}_k)$ is described by three terms:

1. The a priori pdf $p(\mathbf{x}_k|\mathcal{Y}_{k-1})$ defines the knowledge of the model

$$p(\mathbf{x}_k|\mathcal{Y}_{k-1}) = \int p(\mathbf{x}_k|\mathbf{x}_{k-1})p(\mathbf{x}_{k-1}|\mathcal{Y}_{k-1})d\mathbf{x}_{k-1}, \tag{4.2}$$

   where $p(\mathbf{x}_k|\mathbf{x}_{k-1})$ is the transition density of the state.
2. The likelihood $p(\mathbf{y}_k|\mathbf{x}_k)$ essentially determines the measurement noise model in the equation
3. The integral of normalization involving the denominator, also called evidence

$$p(\mathbf{y}_k|\mathcal{Y}_{k-1}) = \int p(\mathbf{x}_k|\mathbf{x}_k)p(\mathbf{x}_k|\mathcal{Y}_{k-1})d\mathbf{x}_k. \tag{4.3}$$

The aim of Bayesian filtering is to apply the Bayesian statistics and Bayesian rule to probabilistic inference problems, and the stochastic filtering problem. Bayesian filtering is optimal in a sense that it seeks the posterior distribution which integrates and uses all of available information expressed by probabilities (assuming they are quantitatively correct). The criteria for measuring the optimality can be done by:

1. The Minimum mean-squared error (MMSE) that is defined in terms of prediction or filtering error (or equivalently the trace of state-error covariance)

$$E[||\mathbf{x}_k - \hat{\mathbf{x}}_k||^2|\mathbf{y}_{0:k}] = \int ||\mathbf{x}_k - \hat{\mathbf{x}}_k||^2 p(\mathbf{x}_k|\mathbf{y}_{0:k})d\mathbf{x}_k \tag{4.4}$$

   where the conditional mean $\hat{\mathbf{x}}_k = E[\mathbf{x}_k|\mathbf{y}_{0:k}] = \int \mathbf{x}_k p(\mathbf{x}_k|\mathbf{y}_{0:k})d\mathbf{x}_k$.
2. Maximum a posteriori (MAP) that is aimed to find the *mode* of posterior probability $p(\mathbf{x}_k|\mathbf{y}_{0:k})$ which is equal to minimize a cost function
3. The Maximum likelihood (ML) which reduces to a special case of MAP where the prior is neglected.
4. The Minimax that is to find the median of posterior $p(\mathbf{x}_k|\mathbf{y}_{0:k})$.

When the *mode* and the *mean* of distribution coincide, the MAP estimation is correct. However, for multimodal distributions, the MAP estimate can be arbitrarily negative. MMSE requires full knowledge of the prior, likelihood and evidence, while MAP methods require the estimation of the posterior distribution (density), but it does not require the calculation of the denominator and thereby is computational inexpensive. Note that, however, MAP estimate has a drawback especially in a high-dimensional space.

4.1.1 The Kalman Filter

Kalman filter can be derived within a Bayesian framework, or more specifically, it reduces to a MAP solution and can be easily also extended to ML solution. Let's consider the following generic stochastic filtering problem in a dynamic state-space form:

$$\dot{\mathbf{x}}_k = \mathbf{f}(k, \mathbf{x}_k, \mathbf{u}_k, \mathbf{w}_k) \tag{4.5}$$

$$\mathbf{y}_k = \mathbf{h}(k, \mathbf{x}_k, \mathbf{u}_k, \mathbf{v}_k), \tag{4.6}$$

where Eqs. (4.5) and (4.6) are called state equation and measurement equation, respectively; $\mathbf{x}_t$ represents the state vector, $\mathbf{y}_k$ is the measurement vector, $\mathbf{u}_k$ represents the system input vector (as driving force) in a controlled environment. $\mathbf{f}$ and $\mathbf{h}$ are two vector valued functions, which are potentially time-varying; $\mathbf{w}_k$ and $\mathbf{v}_k$ represent the process (dynamical) noise and measurement noise respectively, with appropriate dimensions. Since the extension to a driven system is straightforward, the driving force, also called stochastic control problem, is not considered in our approach.

When $\mathbf{F}_k \equiv \frac{\partial \mathbf{f}_k(\mathbf{x})}{\partial \mathbf{x}}$ and $\mathbf{H}_k \equiv \frac{\partial \mathbf{h}_k(\mathbf{x})}{\partial \mathbf{x}}$, the Eqs. (4.5) and (4.6) reduce to the following special case where a linear Gaussian dynamic system is considered:

$$\mathbf{x}_k^t = \mathbf{F}_{k-1}\mathbf{x}_{k-1}^t + \mathbf{w}_{k-1}^t \tag{4.7}$$

$$\mathbf{y}_k = \mathbf{H}_k\mathbf{x}_k^t + \mathbf{v}_k, \tag{4.8}$$

where the true state $\mathbf{x}_k^t$ that we want to estimate has a probability density function $p(\mathbf{x}_k^t)$. Although the probability density function is not available, it is the complete solution of the prediction problem.

Given a set of realizations of all observations available up the time t_l

$$\mathcal{Y}_l^o \equiv \{\mathbf{y}_1^o, \mathbf{y}_2^o, \mathbf{y}_3^o \dots \mathbf{y}_l\}, \tag{4.9}$$

the conditional probability density function given by $p(\mathbf{x}_k^t|\mathcal{Y}_l^o)$ yields to the solution of the filtering problem at time $t_k, k = 1, 2, 3 \dots$. The density probability function $p(\mathbf{x}_k^t|\mathcal{Y}_l^o)$ for fixed l and $t_k, k = l+1, l+2, l+3 \dots$ gives the solution for the prediction problem from time t_1 to t_l.

However in several problems it is not possible compute them because, unlike the unconditional densities, the conditional densities are random functions because they depend on the observations.

Thus, instead of computing the densities explicitly we can use the conditional mean $\hat{\mathbf{x}}_{k|k-1} = E[p(\mathbf{x}_k^t|\mathcal{Y}_k^o)]$ that is a random n-vector since it depends on the realizations of the observations.

The filtering solution is given by the Kalman filter, in which the sufficient statistics of mean and state-error correlation matrix are calculated and propagated. In Eqs. (4.7) and (4.8), $\mathbf{F}_{k-1,k}$ and $\mathbf{H}_k$ are called transition matrix and measurement matrix, respectively. Kalman filter is also optimal in the sense that it is unbiased $E[\hat{\mathbf{x}}_k] = E[\mathbf{x}_k]$ and is a minimum variance estimate. The continuous-time domain, in practice however, is more concerned about the discrete-time filtering, because the continuous-time dynamic system can be always converted into a discrete-time system by sampling the outputs and using *zero-order* holds on the inputs. Hence the derivative will be replaced by the difference and the operator will become a matrix. $\mathbf{w}_k$ and $\mathbf{v}_n$ can be viewed as white noise random sequences with unknown statistics in the discrete-time domain. The state equation characterizes the state transition probability $p(\mathbf{x}_{k+1}|\mathbf{x}_k)$, whereas the measurement equation describes the probability $p(\mathbf{y}_k|\mathbf{x}_k)$ which is related to the measurement noise model. For simplicity we assume the dynamic and measurement noises are both Gaussian distributed with zero mean and constant covariance. The derivation of Kalman filter in the linear Gaussian scenario is based on the following assumptions:

$\mathbf{w}_k^t$ is Gaussian with zero mean and white noise in time, i.e.:

$$\mathbf{w}_k^t \approx N(0, \mathbf{Q}_k); \tag{4.10}$$

$\mathbf{v}_k$ is Gaussian with zero mean and white noise in time, i.e.:

$$\mathbf{v}_k \approx N(0, \mathbf{R}_k); \tag{4.11}$$

$\mathbf{x}_k^t$ is Gaussian with mean $\hat{\mathbf{x}}_k^t$ and covariance $\mathbf{P}_0$, i.e.

$$\mathbf{x}_k^t \approx N(\hat{\mathbf{x}}_k^t, \mathbf{P}_0); \tag{4.12}$$

along with the whiteness assumptions

$$E[\mathbf{w}_k^t(\mathbf{w}_l^t)^T] = \mathbf{Q}_k\delta_{k,l} \quad for \quad all \quad k, l \tag{4.13}$$

$$E[\mathbf{v}_k(\mathbf{v}_l)^T] = \mathbf{R}_k\delta_{k,l} \quad for \quad all \quad k, l \tag{4.14}$$

$$E[\mathbf{w}_k^t(\mathbf{x}_0^t)^T] = E[\mathbf{v}_k(\mathbf{x}_0^t)^T] = E[\mathbf{w}_k^t(\mathbf{v}_k)^T] = 0 \tag{4.15}$$

Suppose we are given the conditional density $p(\mathbf{x}_{k-1}^t|\mathcal{Y}_{k-1}^o)$,the object is to calculate $p(\mathbf{x}_k^t|\mathcal{Y}_k^o)$.

Suppose we are given the conditional density $p(\mathbf{x}_{k-1}^t|\mathcal{Y}_{k-1}^o)$,the object is to calculate $p(\mathbf{x}_k^t|\mathcal{Y}_k^o)$.

4.1.2 The Forecast Step

Denote the mean and covariance matrix respectively of the density $p(\mathbf{x}_{k-1}^t|\mathcal{Y}_{k-1}^o)$

$$\mathbf{x}_{k-1}^a \equiv E[p(\mathbf{x}_{k-1}^t|\mathcal{Y}_{k-1}^o)] \tag{4.16}$$

and

$$\mathbf{P}_{k-1}^a \equiv E[(\mathbf{x}_{k-1}^t - \mathbf{x}_{k-1}^a)(\mathbf{x}_{k-1}^t - \mathbf{x}_{k-1}^a)^T|\mathcal{Y}_{k-1}^o]. \tag{4.17}$$

$\mathbf{x}_{k-1}^a$ is the analysis at time t_{k-1}, an n-vector, the expected value of the true state $\mathbf{x}_k^t$ conditioned on all observations available up to and including that time, while $\mathbf{P}_{k-1}^a$ is the analysis error covariance matrix, an $n \times n$matrix, at time t_{k-1}. At time t there are no observations so from it follows that $p(\mathbf{x}_0^t)$ is a Gaussian density with mean $\mathbf{x}_0^a \equiv \hat{\mathbf{x}}_0^t$ and covariance matrix $\mathbf{P}_0^a \equiv \mathbf{P}_0$ We will see that if $p(\mathbf{x}_{k-1}^t|\mathcal{Y}_{k-1}^o)$ is Gaussian then is $p(\mathbf{x}_k^t|\mathcal{Y}_k^o)$ so by induction it will follow that all the densities $p(\mathbf{x}_1^t|\mathcal{Y}_1^o)$, $p(\mathbf{x}_2^t|\mathcal{Y}_2^o)$... are in fact Gaussian.

In the forecast step denote the mean and covariance matrix of the density $p(\mathbf{x}_k^t|\mathcal{Y}_{k-1}^o)$

$$\mathbf{x}_k^f \equiv E[\mathbf{x}_k^t|\mathcal{Y}_{k-1}^o] \tag{4.18}$$

and

$$\mathbf{P}_k^f \equiv E[(\mathbf{x}_k^t - \mathbf{x}_k^f)(\mathbf{x}_k^t - \mathbf{x}_k^f)^T|\mathcal{Y}_{k-1}^o]. \tag{4.19}$$

$\mathbf{x}_{k-1}^f$ is the forecast at the new time t_{k-1}, an n-vector, the expected value of the true state $\mathbf{x}_k^t$ conditioned on all observations available up to and including that time, while $\mathbf{P}_k^f$ is the forecast error covariance matrix, an $n \times n$ matrix, at time t_k

Substituting (4.7) into (4.18) we obtain:

$$\mathbf{x}_k^f = \mathbf{F}_{k-1} E[\mathbf{x}_{k-1}^t | \mathcal{Y}_{k-1}^o] + E[\mathbf{w}_{k-1}^t | \mathcal{Y}_{k-1}^o], \tag{4.20}$$

since $\mathbf{F}_{k-1}$ is constant, that is a deterministic matrix.

The first expectation here is $\mathbf{x}_{k-1}^a$ along to definition (4.16). The second one is the unconditional expectation $E[\mathbf{w}_{k-1}^t]$ according to the whiteness assumptions (4.15) and therefore vanishes under the assumption which is indeed a forecast to time t_{k-1} via the linear propagator $\mathbf{F}_{k-1}$. Thus we have

$$\mathbf{x}_k^f = \mathbf{F}_{k-1} \mathbf{x}_{k-1}^a, \tag{4.21}$$

which is indeed a forecast to time t_k from the analysis at time t_{k-1} via the linear propagator $\mathbf{F}_{k-1}$.

Substituting (4.7) and (4.21) into (4.19) we obtain:

$$\mathbf{P}_k^f = E[\{\mathbf{F}_{k-1}(\mathbf{x}_{k-1}^t - \mathbf{x}_{k-1}^a) + \mathbf{w}_{k-1}^t\}\{\mathbf{F}_{k-1}(\mathbf{x}_{k-1}^t - \mathbf{X}_{k-1}^a) + \mathbf{w}_{k-1}^t\}^T | \mathcal{Y}_{k-1}^o], \tag{4.22}$$

that under the assumptions (4.15) gives:

$$\mathbf{P}_k^f = \mathbf{F}_{k-1} \mathbf{P}_{k-1}^a \mathbf{F}_{k-1}^T + \mathbf{Q}_{k-1}. \tag{4.23}$$

This equation gives the evolution of $\mathbf{P}_k^f$ starting from $\mathbf{P}_{k-1}^a$. This process does not require any Gaussian assumptions. However if $p(\mathbf{x}_{k-1}^t | \mathbf{X}_{k-1}^o)$ is Gaussian also $p(\mathbf{x}_k^t | \mathbf{X}_{k-1}^o)$ is Gaussian since relation (4.7) is a linear combination of Gaussian random vectors.

4.1.3 The Analysis Step

More complex is the analysis step where we need to make the Gaussian assumption. However let us apply the recursive Bayesian estimation underlying the principle of sequential Bayesian filtering. Under the assumptions that the states follow a first-order Markov process $p(\mathbf{x}_k^t | \mathbf{x}_{0:k-1}) = p(\mathbf{x}_k^t | \mathbf{x}_{k-1})$ and the observations are independent of the given states, we can write:

$$p(\mathbf{x}_k^t | \mathcal{Y}_k^o) = \frac{p(\mathcal{Y}_k^o | \mathbf{x}_k^t) p(\mathbf{x}_k^t)}{p(\mathcal{Y}_k^o)}$$

$$= \frac{p(\mathbf{y}_k, \mathcal{Y}_{k-1}^o|\mathbf{x}_k^t)p(\mathbf{x}_k^t)}{p(\mathbf{y}_k, \mathcal{Y}_{k-1})}$$
$$= \frac{p(\mathbf{y}_k|\mathcal{Y}_{k-1}^o, \mathbf{x}_k^t)p(\mathcal{Y}_{k-1}^o|\mathbf{x}_k^t)p(\mathbf{x}_k^t)}{p(\mathbf{y}_k|\mathcal{Y}_{k-1}^o)p(\mathcal{Y}_{k-1})}$$
$$= \frac{p(\mathbf{y}_k|\mathcal{Y}_{k-1}^o, \mathbf{x}_k^t)p(\mathbf{x}_k^t|\mathcal{Y}_{k-1}^o)p(\mathcal{Y}_{k-1}^o)p(\mathbf{x}_k^t)}{p(\mathbf{y}_k|\mathcal{Y}_{k-1}^o)p(\mathcal{Y}_{k-1}^o)p(\mathbf{x}_k^t)}$$
$$= \frac{p(\mathbf{y}_k|\mathbf{x}_k^t)p(\mathbf{x}_k^t|\mathcal{Y}_{k-1}^o)}{p(\mathbf{y}_k|\mathcal{Y}_{k-1}^o)}. \tag{4.24}$$

We need to evaluate the a posteriori density by the right side of (4.24) that is described by

1. the a priori pdf $p(\mathbf{x}_k^t|\mathcal{Y}_{k-1}^o)$ that defines the knowledge of the model

$$p(\mathbf{x}_k^t|\mathcal{Y}_{k-1}^o) = \int p(\mathbf{x}_k^t|\mathbf{x}_{k-1})p(\mathbf{x}_{k-1}|\mathcal{Y}_{k-1}^o)d\mathbf{x}_{k-1}, \tag{4.25}$$

   where $p(\mathbf{x}_k^t|\mathbf{x}_{k-1})$ is the transition density of the state.
2. the likelihood $p(\mathbf{y}_k|\mathbf{x}_k^t)$ that essentially determines the measurement noise model in the equation
3. the evidence that involves an integral of normalization at the denominator.

$$p(\mathbf{y}_k|\mathcal{Y}_{k-1}^o) = \int p(\mathbf{y}_k|\mathbf{x}_k^t)p(\mathbf{y}_k|\mathcal{Y}_{k-1}^o)d\mathbf{x}_k^t. \tag{4.26}$$

When the *mode* and the *mean* of distribution coincide, the MAP estimation is correct; however, for multimodal distributions, the MAP estimate can be arbitrarily bad. Both MMSE and MAP methods require the estimation of the posterior distribution (density), but MAP does not require the calculation of the denominator (integration) and thereby more computational inexpensive; whereas the former requires full knowledge of the prior, likelihood and evidence. Note that however, MAP estimate has a drawback especially in a high-dimensional space.

Let $\hat{\mathbf{x}}_k^{MAP}$ denote the MAP estimate of $\mathbf{x}_k^t$ that maximizes $p(\mathbf{x}_k^t|\mathcal{Y}_k)$, or equivalently $\log p(\mathbf{x}_k^t|\mathcal{Y}_k^o)$. By using the Bayesian rule, we may express $p(\mathbf{x}_k^t|\mathcal{Y}_k^o)$ by

$$p(\mathbf{x}_k^t|\mathcal{Y}_k^o) = \frac{p(\mathbf{x}_k^t|\mathcal{Y}_k^o)}{p(\mathcal{Y}_k^o)}$$
$$= \frac{p(\mathbf{x}_k^t, \mathbf{y}_k, \mathcal{Y}_{k-1}^o)}{p(\mathbf{y}_k, \mathcal{Y}_{k-1}^o)}, \tag{4.27}$$

where the expression of joint pdf in the numerator is further expressed by

$$p(\mathbf{x}_k^t, \mathbf{y}_k, \mathcal{Y}_{k-1}^o) = p(\mathbf{y}_k|\mathbf{x}_k^t, \mathcal{Y}_{k-1}^o)p(\mathbf{x}_k^t, \mathcal{Y}_{k-1}^o)$$
$$= p(\mathbf{y}_k|\mathbf{x}_k^t, \mathcal{Y}_{k-1})p(\mathbf{x}_k^t|\mathcal{Y}_{k-1}^o)p(\mathcal{Y}_{k-1}^o)$$

and on the fact that $\mathbf{w}_k$ does not depend on $\mathcal{Y}_{k-1}^o$.

$$p(\mathbf{x}_k^t|\mathcal{Y}_k^o) = p(\mathbf{y}_k|\mathbf{x}_k^t)p(\mathbf{x}_k^t|\mathcal{Y}_{k-1}^o)p(\mathcal{Y}_{k-1}^o). \tag{4.28}$$

Substituting (4.28) into (4.27), we obtain:

$$\begin{aligned} p(\mathbf{x}_k^t|\mathcal{Y}_k^o) &= \frac{p(\mathbf{y}_k|\mathbf{x}_k^t)p(\mathbf{x}_k^t|\mathcal{Y}_{k-1}^o)p(\mathcal{Y}_{k-1}^o)}{p(\mathbf{y}_k, \mathcal{Y}_{k-1}^o)} \\ &= \frac{p(\mathbf{y}_k|\mathbf{x}_k^t)p(\mathbf{x}_k^t|\mathcal{Y}_{k-1}^o)p(\mathcal{Y}_{k-1}^o)}{p(\mathbf{y}_k|\mathcal{Y}_{k-1}^o)p(\mathcal{Y}_{k-1}^o)} \\ &= \frac{p(\mathbf{y}_k|\mathbf{x}_k^t)p(\mathbf{x}_k^t|\mathcal{Y}_{k-1}^o)}{p(\mathbf{y}_k|\mathcal{Y}_{k-1}^o)}, \end{aligned} \tag{4.29}$$

which shares the same form as (4.24). Under the Gaussian assumption of process noise and measurement noise, the mean and covariance of $p(\mathbf{y}_k|\mathbf{x}_k^t)$ are calculated by

$$E[\mathbf{y}_k|\mathbf{x}_k^t] = E[\mathbf{H}_k\mathbf{x}_k^t + \mathbf{w}_k|\mathbf{x}_k^t] = \mathbf{H}_k\mathbf{x}_k^t \tag{4.30}$$

and

$$E[(\mathbf{y}_k - E[\mathbf{y}_k|\mathbf{x}_k^t])(\mathbf{y}_k - E[\mathbf{y}_k|\mathbf{x}_k^t])^T|\mathbf{x}_k^t] = E[\mathbf{v}_k(\mathbf{v}_k)^T] = \mathbf{R}_k. \tag{4.31}$$

The conditional pdf $p(\mathbf{y}_k|\mathbf{x}_k^t)$ can be further written as:

$$p(\mathbf{y}_k|\mathbf{x}_k^t) = A_1 \exp\{-\frac{1}{2}(\mathbf{y}_k - \mathbf{H}_k\mathbf{x}_k^t)^T\mathbf{R}_k^{-1}(\mathbf{y}_k - \mathbf{H}_k\mathbf{x}_k^t)\}, \tag{4.32}$$

where $A_1 = (2\pi)^{-N_\mathbf{y}/2}|\mathbf{R}_k|^{-1/2}$, with $N_{\mathbf{y}_k}$ is the dimension of observations.

Remembering the relation (4.19), the covariance $\mathbf{P}_k^f$, by Gaussian assumption, we may obtain:

$$p(\mathbf{x}_k^t|\mathcal{Y}_{k-1}^o) = A_2 \exp\{-\frac{1}{2}(\mathbf{x}_k^t - \mathbf{x}_k^f)^T(\mathbf{P}_k^f)^{-1}(\mathbf{x}_k^t - \mathbf{x}_k^f)\}, \tag{4.33}$$

where $A_2 = (2\pi)^{-N_\mathbf{x}/2}|\mathbf{P}_k^f|^{-1/2}$ with $N_\mathbf{x}$ is the dimension of state. By substituting Eqs. (4.32) and (4.33) to (4.27), it further follows:

$$p(\mathbf{x}_k^t|\mathbf{x}_k) \propto A \exp\{-\frac{1}{2}(\mathbf{y}_k - \mathbf{H}_k\mathbf{x}_k^t)^T\mathbf{R}_k^{-1}(\mathbf{y}_k - \mathbf{H}_k\mathbf{x}_k^t) - \frac{1}{2}(\mathbf{x}_k^t - \mathbf{x}_k^f)^T(\mathbf{P}_k^f)^{-1}(\mathbf{x}_k^t - \mathbf{x}_k^f)\}, \tag{4.34}$$

where $A = A_1 A_2$ is a constant. Since the denominator in relation (4.29) is a normalizing constant, the relation (4.34) can be regarded as an unnormalized density, for which it does not affect the next derivation.

Since the MAP estimate of the state is defined by the condition

$$\left.\frac{\partial \log p(\mathbf{x}_k^t|\mathbf{x}_k)}{\partial \mathbf{x}_k^t}\right|_{\mathbf{x}_k^t=\hat{\mathbf{x}}^{MAP}} = 0, \tag{4.35}$$

substituting Eq. (4.34) into (4.35) yields:

$$\hat{\mathbf{x}}^{MAP} = (\mathbf{H}_k^T \mathbf{R}_k^{-1} \mathbf{H}_k + (\mathbf{P}_k^f)^{-1})^{-1} \times ((\mathbf{P}_k^f)^{-1} \mathbf{x}_k^f + \mathbf{H}_k^T \mathbf{R}_k^{-1} \mathbf{y}_k). \tag{4.36}$$

By using the lemma of inverse matrix [1] it is simplified as:

$$\hat{\mathbf{x}}_k^{MAP} = \mathbf{x}_k^f + \mathbf{K}_k(\mathbf{y}_k - \mathbf{H}_k \mathbf{x}_k^f) \tag{4.37}$$

where $\mathbf{K}_k$ is the Kalman gain as defined by

$$\mathbf{K}_k = \mathbf{P}_k^f \mathbf{H}_k^T (\mathbf{H}_k \mathbf{P}_k^f \mathbf{H}_k^T + \mathbf{R}_k)^{-1} \tag{4.38}$$

and the covariance matrix

$$\mathbf{P}_k^a = E[(\mathbf{x}_k^t - \hat{\mathbf{x}}_k^{MAP})(\mathbf{x}_k^t - \hat{\mathbf{x}}_k^{MAP})^T | \mathcal{Y}_k^o]. \tag{4.39}$$

Observing that $\mathbf{y}_k = \mathbf{H}_k \mathbf{x}_k^t + \mathbf{v}_k$, the difference $\mathbf{x}_k^t - \hat{\mathbf{x}}_k^{MAP}$ is:

$$\begin{aligned} \mathbf{x}_k^t - \hat{\mathbf{x}}_k^{MAP} &= \mathbf{x}_k^t - \mathbf{x}_k^f - \mathbf{K}_k(\mathbf{y}_k - \mathbf{H}_k \mathbf{x}_k^f) \\ &= (\mathbf{I} - \mathbf{K}_k \mathbf{H}_k)(\mathbf{x}_k^t - \mathbf{x}_k^f) - \mathbf{K}_k \mathbf{v}_k \end{aligned} \tag{4.40}$$

and taking into account the assumptions (4.15) the covariance matrix (4.39) is

$$\mathbf{P}_k^a = (\mathbf{I} - \mathbf{K}_k \mathbf{H}_k) \mathbf{P}_k^f (\mathbf{I} - \mathbf{K}_k \mathbf{H}_k)^t + \mathbf{K}_k \mathbf{R}_k \mathbf{K}_k^T. \tag{4.41}$$

Rearranging the above equation it reduces to:

$$\mathbf{P}_k^a = \mathbf{P}_k^f (\mathbf{I} - \mathbf{K}_k \mathbf{H}_k). \tag{4.42}$$

Thus far, the Kalman filter is completely derived from MAP principle.

---

[1] For $(\mathbf{A} - \mathbf{B}\mathbf{D}^{-1}\mathbf{C})^{-1}$, it follows from the matrix inverse lemma, or Sherman Morrison Woodbury formula, that is equal to $\mathbf{A}^{-1} + \mathbf{A}^{-1}\mathbf{B}(\mathbf{D} - \mathbf{C}\mathbf{A}^{-1}\mathbf{B})^{-1}\mathbf{C}\mathbf{A}^{-1}$

The same procedure can be extended to Maximum likelihood maximizing the log likelihood

$$\log p(\mathbf{x}_k|\mathcal{Y}_k) = \log p(\mathbf{x}_k, \mathcal{Y}_k) - \log p(\mathcal{Y}_k) \tag{4.43}$$

The optimal estimate is

$$\left.\frac{\partial \log p(\mathbf{x}_k|\mathcal{Y}_k)}{\partial \mathbf{x}_k}\right|_{\mathbf{x}_k=\hat{\mathbf{x}}^{ML}} = 0, \tag{4.44}$$

substituting (4.34) to (4.44), we want to minimize the cost function

$$\mathcal{J} = (\mathbf{y}_k - \mathbf{H}_k\mathbf{x}_k^t)\mathbf{R}_k^{-1}(\mathbf{y}_k - \mathbf{H}_k\mathbf{x}_k^t)^T + (\mathbf{x}_k^t - \hat{\mathbf{x}}^{ML})(\mathbf{P}_k^f)^{-1}(\mathbf{x}_k^t - \hat{\mathbf{x}}^{ML})^T, \tag{4.45}$$

from which we can obtain the same solution for Kalman filter.

We have seen the Kalman filter can be derived within the Bayesian framework. By admitting state-space formulation, Kalman filter provides that the signal process (i.e. state) is regarded as a linear stochastic dynamical system driven by white noise, the optimal filter thus has a stochastic differential structure which makes the recursive estimation possible. Kalman filter is an unbiased minimum variance estimator under Linear Quadratic Gaussian control circumstance. When the Gaussian assumption of noise is violated, Kalman filter is still optimal in a mean square sense, but the estimate does not produce the conditional mean (i.e. it is biased), and neither the minimum variance. Kalman filter is not robust because of the underlying assumption of noise density model.

In order to approximate the optimal estimation methods for linear problems with gaussian statistics, Monte Carlo methods have been used to estimate the initial condition for forecasting (Evensen and van Leuwen [1], Houtemaker and Mitchel [2])

4.1.4 Prediction by Stochastic Filtering

We will formulate the continuous-time stochastic filtering problem by Stochastic Differential Equation (SDE) theory. Suppose (SDE) $\{\mathbf{x}_k\}$ is a Markov process with an infinitesimal generator, rewriting state-space Eqs. (4.5) and (4.6) in the following form of *Itô* SDE:

$$d\mathbf{x}_k = \mathbf{f}(k, \mathbf{x}_k)dt + \sigma(k, \mathbf{x}_k)d\mathbf{w}_k \tag{4.46}$$

$$d\mathbf{y}_k = \mathbf{h}(k, \mathbf{x}_k)dk + d\mathbf{v}_k, \tag{4.47}$$

where $\mathbf{f}(k, \mathbf{x}_k)$ is often called nonlinear drift and $\sigma(t, \mathbf{x}_k)$ is diffusion coefficient. Again, the noise processes $\{\mathbf{w}_k, \mathbf{v}_k, k \geq 0\}$ are two Brownian or Wiener processes. First, let's look at the state diffusion equation. For all $t \geq 0$, we define a backward diffusion operator $\mathbf{L}_k$, a partial differential operator, as:

$$\mathbf{L}_k = \sum_{i=1}^{N_x} \mathbf{f}_k^i \frac{\partial}{\partial x_i} + \frac{1}{2} \sum_{i,j=1}^{N_x} \mathbf{a}_t^{ij} \frac{\partial^2}{\partial \mathbf{x}_i \partial x_j}, \tag{4.48}$$

where $\mathbf{a}_k^{ij} = \sigma^i(k, \mathbf{x}_t)\sigma^j(k, \mathbf{x}_k)$. Operator $\mathbf{L}_k$ corresponds to an infinitesimal generator of the diffusion process $\{\mathbf{x}_k, k \geq 0|\}$. The goal now is to deduce conditions under which one can find a recursive and finite-dimensional (close form) scheme to compute the conditional probability distribution $p(\mathbf{x}_k|\mathcal{Y}_k)$, given the filtration $\mathcal{Y}_k$ produced by the observation process (4.6).

Let's define an innovations process, that is defined as a white Gaussian noise process

$$\mathbf{e}_k = \mathbf{y}_k - \int_0^k E[\mathbf{h}(k, \mathbf{x}_k)|\mathbf{y}_{0:k}]ds, \tag{4.49}$$

where $E[\mathbf{h}(k, \mathbf{x}_k)|\mathcal{Y}_k]$ is described as:

$$\begin{aligned} \hat{\mathbf{h}}(\mathbf{x}_k) &= E[\mathbf{h}(k, \mathbf{x}_k)|\mathcal{Y}_k] \\ &= \int_{\infty}^{\infty} \mathbf{h}(\mathbf{x}_k) p(\mathbf{x}_k|\mathcal{Y}_k) d\mathbf{x}. \end{aligned} \tag{4.50}$$

For any test function Φ, the forward diffusion operator $\tilde{\mathbf{L}}$ is defined as

$$\tilde{\mathbf{L}}_t \Phi = \sum_{i=1}^{N_x} \mathbf{f}_t^i \frac{\partial \Phi}{\partial x_i} + \frac{1}{2} \sum_{i,j=1}^{N} \mathbf{a}_k^{ij} \frac{\partial^2 \Phi}{\partial \mathbf{x}_i \partial x_j}, \tag{4.51}$$

which essentially is the Fokker-Planck operator. Given initial condition $p(\mathbf{x}_0)$ at $k = 0$ as boundary condition, it turns out that the pdf of diffusion process satisfies the Fokker-Planck-Kolmogorov equation (FPK; a.k.a. Kolmogorov forward equation).The stochastic process is determined equivalently by the FPK equation

$$\frac{\partial p(\mathbf{x}_k)}{\partial k} = \tilde{\mathbf{L}}_k p(\mathbf{x}_k) \tag{4.52}$$

or the SDE (4.46).

The FPK equation can be interpreted as follows: the first term is the equation of motion for a cloud of particles whose distribution is $p(\mathbf{x}_k)$, each point of which obeys the equation of motion $\frac{d\mathbf{x}}{dk} = \mathbf{f}(\mathbf{x}_k, k)$. The second term describes the disturbance due to Brownian motion or as it also called, Wiener noise.

The (4.52) can be solved exactly by Fourier transform. By inverting the Fourier transform, we can obtain

$$p(\mathbf{x}, k + \Delta k|\mathbf{x}_0, k) = \frac{1}{\sqrt{2\pi\sigma_0 \Delta k}} \exp\{-\frac{(\mathbf{x} - \mathbf{x}_0 - \mathbf{f}(\mathbf{x}_0)\Delta k)^2}{2\sigma_0 \Delta k}\}, \tag{4.53}$$

which is a Gaussian distribution of a deterministic path.

The Eq. (4.52) can be considered as the fundamental equation for the time evolution of error statistics. It describes the change of the pdf in a local volume that depends on the divergence term describing a probability flux into the local volume. The diffusion term tends to flatten the pdf due to the effect of stochastic model errors, that is the probability decrease and error increases. When, as in the case of (4.53), we are able to solve the pdf we are also able to be calculate the mean and error covariance for the model forecast at different time levels. This is the case of Kalman filter where pdf is fully characterized by its mean and covariance. However when we have a non linear model the time evolution of $\mathbf{\Phi}(\mathbf{x}_{ij})$ is not able to characterize the mean and covariance and the solution is to solve the FPK by using Monte Carlo methods. If the pdf is represented by a large ensemble of model states it is possible to integrate each member of the ensemble forward in time using the stochastic model (4.46), that is equivalent to solving the FPK equation with Monte Carlo.

The standard approach is to calculate firstly the best guess initial condition on the base of the data and statistics. The ensemble of initial states is generated with the mean that is equal to the best initial condition and the variance is specified from the known uncertainties of the first guess initial state. The covariance should reflect the true scale of the system. In order to provide a realistic increase in the ensemble variance the estimate of the model error variance should be include to give the time evolution of the external error in the estimates.

4.2 Ensemble Kalman Filter

The Ensemble Kalman filter (EnKF) was introduced by Evensen [3, 4] in alternative to the Extended Kalman filter (EKF) (see Kalman chapter) and it is a Monte Carlo approximation to the traditional Kalman Filter. The EnKF is a sequential data assimilation method in which the error statistics are predicted by solving the Fokker Planck or forward Kolmogorov equation with Monte Carlo or ensemble integration by which it is possible to compute the statistical moments like mean and error covariance. Respect to the basic KF the gain of the update equation is calculated from the error covariance provide by the ensemble of model states. The ensemble mean is the best estimate and the spreading of the ensemble around the mean. If we assume to have N model states in the ensemble, each of dimension n, they can be represented as a single point in an n-dimensional state space. In the state space the ensemble members will constitute a cloud points that is described by a probability density function of the type $f(x) = \frac{dn}{n}$ where dn is the number of points in a small unit volume and n is the total number of points. When we know the ensemble representing $f(x)$ we can calculate the moments of the statistics.

Thus the Ensemble Kalman Filter seeks to mimic the Kalman Filter but with an ensemble of limited size rather than with error covariance matrix. The aim of EnKF is to perform an analysis through each member of the ensemble

$$\mathbf{x}_i^a = \mathbf{x}_i^f + \mathbf{K}^{EnKF}(\mathbf{y}_i - \mathbf{H}(\mathbf{x}_i^f)), \tag{4.54}$$

where $i = 1, 2, 3, \ldots n$ is the index indicating the member of the ensemble and $\mathbf{x}_i^f$ is the state vector describing the i member forecasted at the time analysis. $\mathbf{K}^{EnKF}$ should be the Kalman gain we should compute from the ensemble statistics.

$$\mathbf{K}^{EnKF} = \mathbf{P}^f\mathbf{H}^T(\mathbf{H}\mathbf{P}^f\mathbf{H}^T + \mathbf{R})^{-1}, \tag{4.55}$$

where $\mathbf{R}$ is the given observational error covariance matrix. Following Whitaker and Hamill [5]) the forecasted error covariance matrix is estimated from the ensemble

$$\mathbf{P}^f = \frac{1}{n-1}\sum_{i=1}^{n}(\mathbf{x}^f - \bar{\mathbf{x}}^f)(\mathbf{x}^f - \bar{\mathbf{x}}^f)^T, \tag{4.56}$$

with

$$\bar{\mathbf{x}}^f = \frac{1}{n}\sum_{i=1}^{n}\mathbf{x}_i^f. \tag{4.57}$$

Note that wherever an overbar is used in the context of a covariance estimate a factor of $n-1$ instead of n is at denominator that implies the estimate is unbiased.

Thanks to (4.54) we can obtain a posterior ensemble$\{\mathbf{x}_1^a\}$ with $i = 1, 2, 3, \ldots n$ from which we can compute the posterior statistics. Thus the analysis is computed as the mean of the posterior ensemble:

$$\bar{\mathbf{x}}^a = \frac{1}{n}\sum_{i=1}^{n}\mathbf{x}_i^a. \tag{4.58}$$

If all members are updated with the same observations $\mathbf{y}_i \equiv \mathbf{y}$ the deviations of the ensemble members from the mean, that represent the ensemble anomalies $\mathbf{e}^a = \mathbf{x}_i^a - \bar{\mathbf{x}}^a$ implies:

$$\mathbf{e}_i^a = \mathbf{e}_i^f + \mathbf{K}^{EnKF}(\mathbf{0} - \mathbf{H}\mathbf{e}_i^f) = (\mathbf{I}_n - \mathbf{K}^{EnKF}\mathbf{H})\mathbf{e}_i^f, \tag{4.59}$$

that applied to the analysis error covariance matrix leads:

$$\mathbf{P}^a = \frac{1}{n}\sum_{i=1}^{n}(\mathbf{e}^a = \mathbf{x}_i^a - \bar{\mathbf{x}}^a)(\mathbf{e}^a = \mathbf{x}_i^a - \bar{\mathbf{x}}^a)^T = (\mathbf{I}_n - \mathbf{K}^{EnKF}\mathbf{H})\mathbf{P}^f(\mathbf{I}_n - \mathbf{K}^{EnKF}\mathbf{H})^T, \tag{4.60}$$

where $\mathbf{K}^{EnKF} \equiv \mathbf{K}$ Comparing this equation with (4.41) shows the second term is missed and the underestimation can lead to the divergence of EnKF.

A solution can be that to perturb the observation vector for each member: $\mathbf{y}_i = \mathbf{y}+\mathbf{u}_i$ where $\mathbf{u}_i$ is a Gaussian distribution of the type $\mathbf{u}_i \sim N(0, \mathbf{R})$. Let's now define an empirical error covariance matrix

$$\mathbf{R}_u = \frac{1}{n-1}\sum_{i=1}^{n} \mathbf{u}_i \mathbf{u}_i^T, \tag{4.61}$$

where in order to avoid biases $\sum_{i=1}^{n} \mathbf{u}_i = 0$. $\mathbf{R}_u \equiv \mathbf{R}$ in the symptotic limit $n \rightarrow \infty$ With the introduction of this perturbation we modify the anomalies accordingly

$$\mathbf{e}_i^a = \mathbf{e}_i^f + \mathbf{K}_u^{EnKF}(\mathbf{e}_i^o - \mathbf{H}\mathbf{e}_i^f), \tag{4.62}$$

that yields the corrected analysis error covariance matrix

$$\mathbf{P}^a = (\mathbf{I}_n - \mathbf{K}_u^{EnKF}\mathbf{H})\mathbf{P}^f(\mathbf{I}_n - \mathbf{K}_u^{EnKF}\mathbf{H})^T + \mathbf{K}_u^{EnKF}\mathbf{R}_u(\mathbf{K}_u^{EnKF})^T = (\mathbf{I}_n - \mathbf{K}_u^{EnKF}\mathbf{H})\mathbf{P}^f. \tag{4.63}$$

Also in the forecast step the updated ensemble obtained at the analysis evolves in time and is propagated according to a model

$$\mathbf{x}_i^f = \mathbf{M}(\mathbf{x}_i^a) \quad for \quad i = 1, 2, n \tag{4.64}$$

where $\mathbf{M}$ is a model operator. The forecast estimate is the mean of the forecasted ensemble $\bar{\mathbf{x}}^f = \frac{1}{n}\sum_{i=1}^{n} \mathbf{x}_i^f$, while the forecast error covariance matrix is

$$\mathbf{P}^f = \frac{1}{n-1}\sum_{i=1}^{n}(\mathbf{x}_i^f - \bar{\mathbf{x}}^f)(\mathbf{x}_i^f - \bar{\mathbf{x}}^f)^T. \tag{4.65}$$

The most important difference between ensemble Kalman filtering and the other methods is that it only quantifies uncertainty in the space spanned by the ensemble. We can be under a severe limitation if computational resources restrict the number of ensemble members n to be much smaller than the number of model variables m. When this limitation is overcome, then the analysis can be performed in a much lower-dimensional space (n versus m). Thus, ensemble Kalman filtering has the potential to be more computationally efficient than the other methods.

4.2.1 The Stochastic Ensemble Kalman Filter Menu

1. Initialization
   - Initial system state $\mathbf{x}_0^f$ and initial covariance matrix $\mathbf{P}_0^f$

2. For $t_k = 1, 2 \ldots$

   a. Observation

      - Draw a statistical consistent observation set for $i = 1, 2, \ldots n$:

$$\mathbf{y}_i = \mathbf{y} + \mathbf{u}_i \quad \sum_i^n \mathbf{u}_i = 0 \tag{4.66}$$

      - Compute the empirical error covariance matrix

$$\mathbf{R}_u = \frac{1}{n-1} \sum_i^n \mathbf{u}_i \mathbf{u}_i^T \tag{4.67}$$

3. Analysis

   - Compute the gain

$$\mathbf{K}_u = \mathbf{P}^f \mathbf{H}^T (\mathbf{H}\mathbf{P}^f \mathbf{H}^T + \mathbf{R}_u)^{-1} \tag{4.68}$$

   - Ensemble of analysis for $i = 1, 2 \ldots, n$

$$\mathbf{x}_i^a = \mathbf{x}_i^f + \mathbf{K}_u(\mathbf{y}_i - \mathbf{H}(\mathbf{x}_i^f)) \tag{4.69}$$

   and their mean

$$\bar{\mathbf{x}}_i^a = \frac{1}{n} \sum_i^n \mathbf{x}_i^a \tag{4.70}$$

   - Compute the analysis error covariance matrix

$$\mathbf{P}^a = \frac{1}{n-1} \sum_i^n (\mathbf{x}_i^a - \bar{\mathbf{x}}_i^a)(\mathbf{x}_i^a - \bar{\mathbf{x}}_i^a)^T \tag{4.71}$$

4. Forecast

   - Compute the ensemble forecast for $i = 1, 2, \ldots, n$, $\mathbf{x}_i^f = \mathbf{M}(\mathbf{x}_i^a)$ and their mean

$$\bar{\mathbf{x}}_i^f = \frac{1}{n} \sum_i^n \mathbf{x}_i^f \tag{4.72}$$

   - Compute the forecast error covariance matrix

$$\mathbf{P}^f = \frac{1}{n-1} \sum_i^n (\mathbf{x}_i^f - \bar{\mathbf{x}}_i^f)(\mathbf{x}_i^f - \bar{\mathbf{x}}_i^f)^T \tag{4.73}$$

4.2.2 The Deterministic Ensemble Kalman Filter

Let's now define a revised EnKF in which we eliminate the necessity to introduce the perturbation on the observations. There are several variants of the deterministic EnKF, (Whitaker and Hamill [5], Bishop et al [6], Hunt et al. [7]). We report here the one developed by Bishop et al. [6], Hunt et al. [7] that can be defined as Ensemble Transform (ETKF) or one of the variant of the Ensemble square root Kalman Filter (EnSRKF). The other approaches (EnSRKF and LETKF) are presented in the applications' chapter, where also presented is the Particle filters (Monte Carlo) method applied to a highly non-Gaussian distribution and non linear operator.

The entire approach is based on the ensemble space rather than in the state or observation space. We start with an ensemble $\{(\mathbf{X}_i^f)_k\}$ $i = 1, 2, \ldots, n$ of n-dimensional model state vectors at time t_{k-1}. Rather than to let one of the ensemble members represent the best estimate of the system state, one assumes the ensemble to be chosen so that its average represents the analysis state estimate. We evolve each ensemble member according to the nonlinear model to obtain a forecast ensemble $\{(\mathbf{X}_i^f)_{k-1},\quad i = 1, 2, \ldots, n\}$ at time t_k:

$$[\mathbf{X}_i^f]_k = \mathbf{M}_{k-1,k}([\mathbf{X}_i^a]_{k-1}). \tag{4.74}$$

Let l the number of scalar observations used in the analysis. Starting from the sample mean $\bar{\mathbf{x}}^f = \frac{1}{n}\sum_{i=1}^n \mathbf{X}_i^f$ the forecast error covariance matrix (4.65) can be written:

$$\mathbf{P}^f = \frac{1}{n-1}\sum_{i=1}^{n}(\mathbf{X}_i^f - \bar{\mathbf{x}}^f)(\mathbf{x}_i^f - \bar{\mathbf{x}}^f)^T = \mathbf{X}^f(\mathbf{X}^f)^T, \tag{4.75}$$

where $\mathbf{X}^f$ is a $m \times n$ matrix composed by n columns m-vectors whose columns are the normalized anomalies:

$$[\mathbf{X}^f]_i = \frac{\mathbf{x}_i^f - \bar{\mathbf{x}}^f}{\sqrt{n-1}}. \tag{4.76}$$

The analysis not only determines and estimate, a covariance but also an ensemble $\{(\mathbf{x}_i^a)_{k-1}\}$, $i = 1, 2, \ldots, n$ with the appropriate sample mean and covariance.

$$\bar{\mathbf{x}}^a = \frac{1}{n}\sum_{i=1}^{n}\mathbf{x}_i^a \tag{4.77}$$

and

$$\mathbf{P}^a = \frac{1}{n-1}\sum_{i=1}^{n}(\mathbf{x}_i^a - \bar{\mathbf{x}}^a)(\mathbf{x}_i^a - \bar{\mathbf{x}}^a)^T = \mathbf{X}^a(\mathbf{X}^a)^T, \tag{4.78}$$

where $\mathbf{x}^a$ is the $m \times n$ whose columns are

$$[\mathbf{X}^a]_i = \frac{\mathbf{x}_i^a - \bar{\mathbf{x}}^a}{\sqrt{n-1}}. \tag{4.79}$$

4.2.3 The Analysis Scheme

In order to describe the transformation from a forecast ensemble $\{(\mathbf{X}_i^f)_k, \quad i = 1, 2, \ldots, n\}$ into an appropriate analysis ensemble $\{(\mathbf{X}_i^a)_{k-1}, \quad i = 1, 2, \ldots, n\}$ we assume that the number of ensemble members n is smaller than both the number of model variables m and the number of observations l, even when localization has reduced the effective values of m and l considerably compared to a global analysis. Assuming the choice of observations to use for the local analysis to have been performed already, and consider $\mathbf{y}$, $\mathbf{H}$ and $\mathbf{R}$ to be truncated to these observations; as such, correlations between errors in the chosen observations and errors in other observations are ignored. Most of the analysis will take place in a n-dimensional space, with as few operations as possible in the model and observation spaces.

Remember the relation (4.45) where we want to minimize the cost function $\mathcal{J}$ where we have the forecast covariance matrix having rank at most $n-1$ and therefore it is not invertible. Nevertheless, its inverse is well-defined on the space $\mathcal{S}$ spanned by the background ensemble perturbations, that is, the columns of $\mathbf{x}^f$. Thus $\mathcal{J}$ is also well-defined in S, and the minimization can be carried out in this subspace. The reduced dimensionality is an advantage from the point of view of efficiency, though the restriction of the analysis mean to S is sure to be detrimental if n is too small. A natural approach is to use the singular vectors of $\mathcal{Y}^f$ (the eigenvectors of $\mathbf{P}^f$) to form a basis for S. In order to perform the analysis on S, we must choose an appropriate coordinate system even though a conceptual difficulty in this approach is that the sum of these columns is zero, so they are necessarily not linearly independent. We could assume the first $n-1$ columns to be independent and use them as a basis, but this assumption is unnecessary and clutters the resulting equations. Instead, we regard $\mathbf{x}^f$ as a linear transformation from a n-dimensional space $\tilde{\mathcal{S}}$ onto $\mathcal{S}$, and perform the analysis in $\tilde{\mathcal{S}}$. Let $\mathbf{d}$ denote a vector in $\tilde{\mathcal{S}}$ then $\mathbf{X}^f\mathbf{d}$ belongs to the space $\mathcal{S}$ spanned by the background ensemble perturbations, and

$$\mathbf{x} = \bar{\mathbf{x}}^f + \mathbf{X}^f\mathbf{d} \tag{4.80}$$

is the corresponding model state in an affine space. Note that if $\mathbf{d}$ is a Gaussian random vector with mean 0 and covariance $(n-1)^{-1}\mathbf{I}$, relation (4.80) is gaussian with mean $\bar{\mathbf{x}}^f$ and covariance

$$\mathbf{P}^f = \frac{1}{n-1}\sum_{i=1}^{n}(\mathbf{x}_i^f - \bar{\mathbf{x}}^f)(\mathbf{x}_i^f - \bar{\mathbf{x}}^f)^T = \mathbf{X}^f(\mathbf{X}^f)^T. \tag{4.81}$$

If the observation operator is linear we can also use the notation $\mathbf{Y}^f = \mathbf{H}\mathbf{X}^f$. If the operator is not linear we consider $\mathbf{Y}^f$ to be the matrix of the observation anomalies:

$$[\mathbf{Y}^f]_i = \frac{\mathbf{H}(\mathbf{x}_i^f) - \bar{\mathbf{y}}^f}{\sqrt{n-1}}, \tag{4.82}$$

with $\bar{\mathbf{y}} = \frac{1}{n}\sum_{i=1}^{n}\mathbf{H}(\mathbf{x}_i^f)$.

With respect to the stochastic EnKF one performs a single analysis rather than performing an analysis for each member of the ensemble. Adapting the mean analysis of the stochastic EnKF we have:

$$\bar{\mathbf{x}}^a = \bar{\mathbf{x}}^f + \mathbf{K}^*(\mathbf{y} - \mathbf{H}\bar{\mathbf{x}}^f), \tag{4.83}$$

where $\mathbf{K}^* = \mathbf{P}^f\mathbf{H}^T(\mathbf{H}\mathbf{P}^f\mathbf{H}^T + \mathbf{R})^{-1}$ rather than $\mathbf{K}_u$, because in the deterministic approach the observations are not perturbed.

As before we reformulate our analysis in the ensemble space, that is to find the corresponding optimal coefficient vector $\mathbf{d}^a$ that stands for $\mathbf{x}^a$ in our relation:

$$\mathbf{x}^a = \bar{\mathbf{x}}^f + \mathbf{X}^f\mathbf{d}^a. \tag{4.84}$$

Inserting this decomposition into (4.83) we obtain

$$\bar{\mathbf{x}}^f + \mathbf{X}^f\mathbf{d}^a = \bar{\mathbf{x}}^f + \mathbf{X}^f(\mathbf{x}^f)^T\mathbf{H}^T(\mathbf{H}\mathbf{X}^f(\mathbf{X}^f)^T\mathbf{H}^T + \mathbf{R})^{-1}(\mathbf{y} - \mathbf{H}\bar{\mathbf{x}}^f). \tag{4.85}$$

If we remember that $\mathbf{X}^f\mathbf{H} = \mathbf{Y}^f$ simplifying and substituting we obtain:

$$\begin{aligned}\mathbf{d}^a &= (\mathbf{X}^f)^T\mathbf{H}^T(\mathbf{H}\mathbf{X}^f(\mathbf{X}^f)^T\mathbf{H}^T + \mathbf{R})^{-1}(\mathbf{y} - \mathbf{H}(\bar{\mathbf{x}}^f))\\ &= (\mathbf{Y}^f)^T(\mathbf{Y}^f(\mathbf{Y}^f)^T + \mathbf{R})^{-1}(\mathbf{y} - \mathbf{H}(\bar{\mathbf{x}}^f)).\end{aligned} \tag{4.86}$$

Applying the Sherman-Morrison-Woodbury lemma $\mathbf{I}_n(\mathbf{Y}^f)^T(\mathbf{Y}^f(\mathbf{Y}^f)^T + \mathbf{R})^{-1}$ we obtain:

$$\mathbf{d}^a = (\mathbf{I}_n + (\mathbf{Y}^f)^T\mathbf{R}^{-1}\mathbf{Y}^f)(\mathbf{Y}^f)^T\mathbf{R}^{-1}(\mathbf{y} - \mathbf{H}(\bar{\mathbf{x}}^f)). \tag{4.87}$$

Then the gain is computed into the ensemble space rather than in the observation space, that mean we are reformulating the stochastic EnKF.

In order to understand the difference between the deterministic EnKF and the stochastic EnKF we can generate the posterior ensemble. Remembering relation (4.63) we want to factorize $\mathbf{P}^a = \mathbf{X}^a(\mathbf{X}^a)^T$. In fact

$$\begin{aligned}\mathbf{P}^a &= (\mathbf{I}_n - \mathbf{K}_u^{EnKF}\mathbf{H})\mathbf{P}^f\\ &\simeq (\mathbf{I}_n - \mathbf{X}^f(\mathbf{Y}^f)^T(\mathbf{Y}^f(\mathbf{Y}^f)^T + \mathbf{R}^{-1})\mathbf{Y}^f(\mathbf{Y}^f)^T\\ &\simeq \mathbf{X}^f(\mathbf{I}_n - (\mathbf{Y}^f)^T(\mathbf{Y}^f(\mathbf{Y}^f)^T + \mathbf{R})^{-1}\mathbf{Y}^f)(\mathbf{X}^f)^T.\end{aligned} \tag{4.88}$$

Now we are looking for a square root matrix such as $\mathcal{Y}^a(\mathcal{Y}^a)^T = \mathbf{P}^a$. One of this matrix is

$$\mathbf{X}^a = \mathbf{X}^f(\mathbf{I}_n - (\mathbf{Y}^f)^T(\mathbf{Y}^f(\mathbf{Y}^f)^T + \mathbf{R})^{-1}\mathbf{Y}^f)^{\frac{1}{2}} \tag{4.89}$$

that can be simplified using the Sherman-Morrison-Woodbury lemma, obtaining:

$$\mathbf{X}^a = \mathbf{X}^f(\mathbf{I}_n - (\mathbf{Y}^f)^T\mathbf{R}^{-1}\mathbf{Y}^f)^{-\frac{1}{2}}. \tag{4.90}$$

This is the a posteriori ensemble of anomalies for the deterministic EnKF. Defining as $\mathbf{T} = (\mathbf{I}_n + (\mathbf{Y}^f)^T\mathbf{R}^{-1}\mathbf{Y}^f)^{-\frac{1}{2}}$ we can build the posterior ensemble as:

$$\mathbf{x}_i^a = \bar{\mathbf{x}}^a + \sqrt{n-1}\mathbf{X}^f[\mathbf{T}]_i = \bar{\mathbf{x}}^f + \mathbf{X}^f(\mathbf{d}^a + \sqrt{n-1}\mathbf{X}^f[\mathbf{T}]_i). \tag{4.91}$$

We can note the ensemble is centered on $\bar{\mathbf{x}}^a$. In fact

$$\frac{1}{n}\sum_{i=1}^{n}\mathbf{x}_i^a = \bar{\mathbf{x}}^a + \frac{\sqrt{n-1}}{n}\mathbf{X}^f\mathbf{T}\mathbf{1} = \bar{\mathbf{x}}^a + \frac{\sqrt{n-1}}{n}\mathbf{X}^f\mathbf{1} = \bar{\mathbf{x}}^a, \tag{4.92}$$

because $\mathbf{1} = [1, 1, 1, 1..]^T$ is the matrix that yields the same state vector $\mathbf{x}$ since $\mathbf{X}^f\mathbf{1} = \mathbf{0}$.

4.2.4 The Deterministic Ensemble Kalman Filter Menu

1. Initialization
   - Ensemble of state vectors $\mathbf{E}_0^f = \{\mathbf{x}_0, \ldots \mathbf{x}_n\}$
2. For $t_k = 1, 2, \ldots$
   a. Analysis
      - Compute forecast mean, the ensemble anomalies and the observation anomalies:

$$\mathbf{x}_k^f = \frac{1}{n}\mathbf{E}_k^f\mathbf{1}$$
$$\mathbf{Y}_k^f = \frac{1}{\sqrt{n-1}}(\mathbf{E}_k^f - \bar{\mathbf{x}}^f\mathbf{1}^T)$$
$$\mathbf{Y}_k^f = \frac{1}{\sqrt{n-1}}(\mathbf{H}_k(\mathbf{E}_k^f) - \mathbf{H}_k(\bar{\mathbf{x}}^f)\mathbf{1}^T)$$

      - Computation of ensemble transform matrix

$$\mathbf{T} = (\mathbf{I}_n + \mathbf{Y}_k^T\mathbf{R}_k^{-1}\mathbf{Y}_k)^{-1} \tag{4.93}$$

- Analysis estimate in ensemble space

$$\mathbf{d}_k^a = \mathbf{T}\mathbf{Y}_k^T\mathbf{R}^{-1}(\mathbf{y}_k - \mathbf{H}(\bar{\mathbf{x}}_k^f)) \tag{4.94}$$

- Generating the posterior ensemble

$$\mathbf{E}_k^a = \bar{\mathbf{x}}_k^f\mathbf{1}^T + \mathbf{X}_k^f(\mathbf{d}_k^a\mathbf{1}^T + \sqrt{n-1}\mathbf{T}^{\frac{1}{2}}) \tag{4.95}$$

b. Forecast
- Forecast ensemble

$$\mathbf{E}_{k+1}^f = \mathbf{M}_{k+1}(\mathbf{E}_k^a) \tag{4.96}$$

4.3 Issues Due to Small Ensembles

Ensemble Kalman filtering is subjected to some restriction when the computation requires of maintaining a large ensemble. Ensembles that are too small can lead to problems of a different nature. Since the specification of the relative weighting to be placed on the background errors is always underestimated, Furrer and Bengtsson [8] refer this can be exacerbated by using ensembles sizes that are too small. This leads to the filter placing more confidence in the background state of the system and less on the observations. This means that observational data is not influential in adjusting the forecast state adequately when producing the analysis state. If the forecast state is not a good representation of the true state of the system then filter fails to produce a meaningful state estimate. Further the use of a small ensemble can lead to the development of correlations between state components that are at a significant distance from one another where there is no physical relation.

Thus the success of the EnKF depends on the size of the ensemble employed as it has been investigated widely for example in Houtekamer and Mitchell [2]. The success of an ensemble filter will be dependent on the ensemble being statistically representative but the ensemble must span the model sub-space adequately (Oke et al., 2007). Current NWP models have state spaces of the order of $O(10^7)$ (UK Met Office [9]) and thus require a large ensemble to be adequately statistically represented. This requires a large computational cost. If an ensemble filter uses a smaller number of ensemble members than the size of the state, that ensemble is not statistically representative and the system is said to be undersampled introducing three major problems in ensemble filtering: inbreeding, filter divergence and the development of long range spurious correlations.

4.3.1 Inbreeding

Inbreeding has been introduced by Houtekamer and Mitchell [2] to describe the phenomenon in ensemble filtering that arise due to undersampling. This term is also used to describe a situation where the analysis error covariances are systematically underestimated after each of the observation assimilations. Furrer and Bengtsson [8] indicate the analysis error covariance $\mathbf{P}^a$ should always be less than that of the forecast error covariance, since, as we have been before, it is defined as:

$$\mathbf{P}^a = (\mathbf{I}_n - \mathbf{KH})\mathbf{P}^f, \tag{4.97}$$

where the forecast error covariance $\mathbf{P}^f = E[(\mathbf{x}^f - \mathbf{x}^t)(\mathbf{x}^f - \mathbf{x}^t)^T]$ is a measure of uncertainty in the forecast estimate state of the system.

The ensemble Kalman gain, as given by the relation $\mathbf{K}_e = \mathbf{P}_e^f\mathbf{H}^T(\mathbf{HP}_e^f\mathbf{H}^T + \mathbf{R})^{-1}$, where the index e indicates the ensemble, uses a ratio of the error covariance of the forecast background state and the error covariance of the observations to calculate the weight that should be placed on the background state and how much weighting should be given to the observations. Then the forecast state estimate is adjusted by the observations in accordance with the ratio of background and observation covariance matrices in the Kalman gain. Note if either the forecast background errors or observational errors are incorrectly specified then the adjustment of the forecast state will be incorrect.

Inbreeding in small ensembles that do not adequately span the model subspace can occur due to sampling errors in the covariance estimate (Lorenc [10]). The smaller the ensemble is, the greater the degree of undersampling is present and the greater the chance is of underestimated forecast error covariances (Ehrendorfer [11]). In ensemble filters, where each member of the ensemble is updated, at the analysis stage, by the same observations and the same Kalman gain matrix there is a tendency for the filter to underestimate the analysis covariance (Whitaker and Hamill [5]). Ensemble Kalman filters that use perturbed observations, such as the EnKF of Evensen [4], have additional sampling errors in the estimation of the observation error covariances. This in turn makes it more likely that inbreeding will occur (Whitaker and Hamill [5]).

Ensemble Kalman filters that use perturbed observations have additional sampling errors in the estimation of the observation error covariances which means it is likely that the inbreeding occurs (Whitaker and Hamill [5]). In case of Square root filters since observations are not perturbed the inbreeding is negated. Inbreeding is a potential source of filter divergence and the development of spurious long range correlations (Hamill et al. [12]). Undersampling produces a reduced rank representation of the background error covariance matrix and in cases where the undersampling is severe there is a tendency for the variances and covariances to be underestimated.

4.3.2 Filter Divergence

When an incorrectly specified analysis state is unable to be adjusted by observation assimilations to more accurately represent the true state of the system a filter divergence occurs. When the covariances in the forecast estimate become too large then there is little certainty in the forecast estimate state of the system because less weighting is given to the forecast state of the system and more to the observations. On the contrary if the covariances become too small then there is high degree of certainty in the forecast estimate state of the system because, due to the Kalman gain, more weighting is placed on the forecast state of the system and less on the observations. The greater the degree of inbreeding, the more relative weighting a filter assigns to the background state and the more the observational data is ignored during the assimilation. If the standard deviation of the analysis state ensemble is smaller than that of the forecast state ensemble the ensemble members are converging. Due to inbreeding, it becomes impossible for the filter to then adjust an incorrectly specified forecast state to accurately represent the true state and filter divergence has occurred [13].

4.3.3 Spurious Correlations

Spurious correlations happen in the forecast error covariance between state components that are not physically related and they are normally at a significant distance from each other. Where all the observations have an impact on each state variable large long range spurious correlations may develop (Anderson [14]). The consequence of these is that a state variable may be incorrectly impacted by an observation that is physically remote. As the size of the ensemble and the true correlation between state components decreases, so the error in the covariance error estimate, relative to the true correlation, greatly increases (Hamill et al. [12]). In the physical world it is expected that at a distance from a given observation point the true correlation will decrease. In a NWP the size of the error relative to the true correlation at grid points remote from the observation point in the forecast error covariance matrix will therefore be expected to increase (Hamill et al. [12]). It was demonstrated by Hamill et al. [12] that the analysis estimate provided by the EnKF was less accurate when the error in the covariance estimate, known as noise, is greater than the true correlation, known as the the signal. Since the correlations at a distance are expected to be small and the relative error increases with distance, then it is expected that state components distant from the observations have a greater noise to signal ratio. These are long range spurious correlations and they degrade the quality of the analysis estimate. Further Hamill et al. [12] show that the noise to signal ratio is a function of ensemble size. Larger ensembles which more accurately reflect the statistics have less associated noise. Thus the problem of spurious correlations is associated with under-sampling. Lorenc [10] shows that when an ensemble is generated by random sampling from a probability distribution function (pdf), the forecast error covariance will have an error proportional to 1/N, where N is the size of the ensemble.

4.4 Methods to Reduce Problems of Undersampling

Various methods have been implemented in an attempt to negate these problems. Methods known as covariance inflation and covariance localization (Hamill et al. [12]) are the methods implemented in the model previously described.

However these are not the only methods used, alternative methods of covariance localization including a double EnKF (Houtekamer and Mitchell [2]) and the local ensemble Kalman filter (LEKF) (Ott et al. [15], Szunyogh et al. [16]) are also possible. The double EnKF employs parallel ensemble data assimilation cycles. Error covariances are estimated by one ensemble and this is then used to update the other. This can prevent the cycle of systematic inbreeding. The LEKF is a square root filter where the analysis equations are solved exactly in a local subspace spanned by all the ensemble members using all the available observations within the predefined localized space [13]. This filter is explored in more detail in Kalnay et al. [17].

4.4.1 Spatial Localization

If the ensemble has n members, then the forecast covariance matrix $\mathbf{P}^f$ given by (4.75) describes nonzero uncertainty only in the n-dimensional subspace spanned by the ensemble, and a global analysis will allow adjustments to the system state only in this subspace. If the system is high-dimensionally unstable, then forecast errors will grow in directions not accounted for by the ensemble, and these errors will not be corrected by the analysis. On the other hand, in case of a sufficiently small local region, the system may behave like a low-dimensionally unstable system driven by the dynamics in neighboring region (see Hunt [7]).

Localization is generally done either explicitly, considering only the observations from a region surrounding the location of the analysis, or implicitly, by multiplying the entries in $\mathbf{P}^f$ by a distance-dependent function that decays to zero beyond a certain distance, so that observations do not affect the model state beyond that distance. Since the model is discretized, the choice of which observations to use for each grid point depends also on the method adopted, and a good choice will depend both on the particular system being modeled and the size of the ensemble (more ensemble members will generally allow more distant observations to be used gainfully). It is important however to have significant overlap between the observations used for one grid point and the observations used for a neighboring grid point; otherwise the analysis ensemble may change suddenly from one grid point to the next.

The global analysis should not be confined to the n-dimensional ensemble space and instead should explore a much higher dimensional space. However the necessity of spatial localization for space temporally chaotic systems introduce barely correlations between distant locations in the background covariance matrix $\mathbf{P}^f$, at least for short time scales. These spurious correlations randomly affect observations. If the system has a characteristic *correlation distance*. then the analysis should ignore ensemble correlations over greater distances. In this context an important issue in ensemble Kalman filtering of space temporally chaotic systems is spatial localization.

4.4.2 Covariance Inflation

One method to correct the underestimation in the forecast error covariance matrix is the Covariance inflation. It was introduced by Anderson and Anderson [18] with the aim to increase the forecast error covariances by "inflating", for each ensemble member, the deviation of the background error from the ensemble mean by a certain percentage. In fact prior to a new observation being assimilated in any new cycle, the background forecast deviations from the mean are increased by a factor of "inflation", $\mathbf{r}$,

$$\mathbf{x}_i^f \leftarrow \mathbf{r}(\mathbf{x}_i^f - \bar{\mathbf{x}}^f) + \bar{\mathbf{x}}^f, \tag{4.98}$$

where $\leftarrow$ represents the replacement of the previous value.

The inflation factor, $\mathbf{r}$, is normally chosen to be slightly greater than 1.0, but the specification of an optimal inflation factor may vary according to the size of the ensemble (Hamill et al. [12]). Hamill et al. [12], experimentally found, that a 1 % inflation factor was nearly optimal for all numbers of ensemble members even though Whitaker and Hamill [5] found the optimal values of the inflation factors were 7 % for the EnKF and 3 % for EnSRF.

The size of the inflation factor chosen depends on various factors. First of all it will depend on the dynamical model adopted. In fact the size of the inflation factor required may be deeply affected if the dynamical model has significant error growth or decay. Furthermore the size of the inflation factor may also be dependent on the type of ensemble filter used, and the covariance filtering length scale can be chosen as in Whitaker and Hamill [5].

This method of overcoming the problem of inbreeding is commonly used, for example in Hamill et al. (2001); Whitaker and Hamill [5]; Anderson [14] and Oke et al. [19]. Anderson [14] mentions that, when there are local linear physical balances in the system dynamics, a large inflations may break this balance. Inflation factors do not help to correct the problem of long range spurious correlations, for this a more sophisticated approach is required.

4.4.3 Covariance Localization

Houtekamer and Mitchell [2], Hamill et al. [12], Whitaker and Hamill [5] show the covariance localization is a process of cutting off longer range correlations in the error covariances at a specified distance. It is a method of improving the estimate of the forecast error covariance. Since the empirical $\mathbf{P}^f$ is a good approximation, the insufficient rank introduces long range correlations that are the above mentioned spurious correlations. Then the idea is to regularize $\mathbf{P}^f$ by shooting out these long range correlation and increasing the rank of $\mathbf{P}^f$ by multiplying it with short range predefined correlation matrices $\boldsymbol{\rho}$. The pointwise multiplication is ordinarily achieved by applying a Schur product (Schur [20]), also known as the Hadamard product (Horn [21]), denote with a $\circ$, to the forecast error covariance matrix.

A Schur product involves an element-wise product of matrices written as $\mathbf{P}^f \circ \boldsymbol{\rho}$, where $\mathbf{P}^f$ and $\boldsymbol{\rho}$ have the same dimensions. If i is the row index and j is the column index the Schur product is calculated as

$$[\mathbf{P}^f \circ \boldsymbol{\rho}]_{ij} = [\mathbf{P}^f]_{ij}[\boldsymbol{\rho}]_{ij}. \tag{4.99}$$

In mathematics, the Hadamard product (also known as the Schur product or the entrywise product) is a binary operation that takes two matrices of the same dimensions, and produces another matrix where each element ij is the product of elements ij of the original two matrices. It should not be confused with the more common matrix product. It is attributed to, and named after, either French mathematician Jacques Hadamard, or German mathematician Issai Schur. The Hadamard product is associative and distributive, and unlike the matrix product it is also commutative (see Wikipedia)

To achieve covariance localization by Schur product a function, $\boldsymbol{\rho}$, is normally defined to be a correlation function with local support. Local support is a term meaning that the function is only non zero in a small (local) region and is zero elsewhere. The correlation function is commonly taken to be the compactly supported 5th order piecewise rational function as defined in Gaspari and Cohn [22], such that

$$\rho = \begin{cases} -\frac{1}{4}(|z|/c)^5 + \frac{1}{2}(|z|/c)^4 + \frac{5}{8}(|z|/c)^3 - \frac{5}{3}(|z|/c)^2 + 1, & 0 \le |z| \le c \\ \frac{1}{12}(|z|/c)^5 - \frac{1}{2}(|z|/c)^4 + \frac{5}{8}(|z|/c)^3 + \frac{5}{3}(|z|/c)^2 & \\ -5(|z|/c) + 4 + \frac{5}{8}(|z|/c)^3 - \frac{2}{3}(c/|z|) & c \le |z| \le 2c \\ 0 & 2c \le |z|, \end{cases}$$

where z is the Euclidean distance between either the grid points in physical space or the distance between a grid point and the observation location; this is dependent on the implementation. A length scale c is defined such that beyond this the correlation reduces from 1 and at a distance of more than twice the correlation length scale the correlation reduces to zero. The length scale is generally set to be $c = \sqrt{\frac{10}{3l}}$ where l is any chosen cutoff length scale. The factor $\sqrt{\frac{10}{3l}}$ is included to tune the correlation function to be optimal (Lorenc [10]) such that the final localised global average error variance is closet to that of the true probability distribution.

To achieve covariance localization a Schur product is taken between the forecast background error covariance matrix, $\mathbf{P}^f$, calculated from the ensemble, and a correlation function with local support, $\boldsymbol{\rho}$. Remembering the Kalman gain (4.55) we can write:

$$\mathbf{K}_e = (\boldsymbol{\rho} \circ \mathbf{P}_e^f)\mathbf{H}^T(\mathbf{H}(\boldsymbol{\rho} \circ \mathbf{P}_e^f)\mathbf{H}^T + \mathbf{R})^{-1}. \tag{4.100}$$

Since $\boldsymbol{\rho}$ is a covariance matrix and $\mathbf{P}^f$ is a covariance matrix then it can be proved that $\mathbf{P}^f \circ \boldsymbol{\rho}$ is also a covariance matrix (Horn [21]).

The distance at which correlations in the error covariances are cutoff, i.e. reduced to zero, can be defined in two ways. In Hamill et al. [12] it is defined as the Euclidean distance between a grid point and the observation location and in Oke et al. [19] as the Euclidean distance between grid points in physical space. The filtering length scale is of primary importance because must remove spurious correlations while the correlation function that correctly specifies physical correlations are not excessively damped but maintained. If the filter length scale is too long, in order to correctly capture all the dynamical correlations, then many of the spurious correlations may not be removed. Ehrendorfer [11] suggests that "the covariance may be noisy causing an overly adjusted variance deficient ensemble". If the length scale is too short then important dynamical correlations may be lost.

Increasing the rank of the forecast error covariance means increasing the number of degrees of freedom of available for assimilating the observations. In such a case the analysis state should better represent the observations (Ehrendorfer [11]). This is useful when a system is under-sampled as the effective size of the ensemble is increased (Oke et al. [19]). The use of the correlation function, where there is an introduction of many zeros into the error covariance matrices also has the advantage of computational savings in the calculations of the analysis error covariance (Lorenc [10]).

References

1. Evensen, G., van Leeuwen, P.: Assimilation of Geosat altimeter data for the Agulhas current using the ensemble Kalman filter with a quasi geostrophic model. Mon. Weather. Rev. **124**, 85–96 (1996)
2. Houtekamer, P.L., Mitchell, H.L.: Data assimilation using an ensemble Kalman filter technique. Mon. Weather. Rev. **126**, 796–811 (1998)
3. Evensen, G.: Sequential data assimilation with nonlinear quasi-geostrophic model using Monte Carlo methods to forecast error statistics. J. Geophys. Res. **99**, 143–162 (1994)
4. Evensen, G.: Data assimilation: the ensemble kalman filter, 2nd edn, p. 320. Springer, New York (2009)
5. Whitaker, J.S., Hamill, T.M.: Ensemble data assimilation without perturbed observations. Mon. Weather. Rev. **130**, 1913–1924 (2002)
6. Bishop, C.H., Etherton, B.J., Manjundar, S.J.: Adaptive sampling with the ensemble transform filter. Part I: theoretical aspects. Mon. Weather. Rev. **129**, 420–436 (2001)
7. Hunt, B.R., Kostelich, E.J., Szunyogh, I.: Efficient data assimilation for spatiotemporal chaos: a local ensemble transform Kalman filter. Physica D **230**, 112–126 (2007)
8. Furrer, R., Bengtsson, T.: Estimation of high-dimensional prior and posterior covariance matrices in Kalman filter variants. J. Multivar. Anal. **98**, 227–255 (2007)
9. UKMet Office.: Observations. https://www.metoffice.gov.uk/research/nwp/observations/. Last accessed (2008)
10. Lorenc, A.C.: The potential of the ensemble Kalman filter for NWP—a comparison with 4D-VAR. Q. J. R. Meteorol. Soc. **129**, 3183–3203 (2003)
11. Ehrendorfer, M.: A review of issues in ensemble-based Kalman filtering. Meteorol. Z. **16**, 795–818 (2007)

12. Hamill, T., Mullen, S., Snyder, C., Toth, Z., Baumhefner, D.: Ensemble forecasting in the short to medium range: report from a workshop. Bull. Am. Meteorol. Soc. **81**, 2653–2664 (2000)
13. Petrie, R.E.: Localization in the ensemble Kalman Filter. MSc Atmosphere, Ocean and Climate University of Reading (2008)
14. Anderson, J.L.: An Ensemble Adjustment Kalman Filter for data assimilation. Mon. Weather. Rev. **129**, 2884–2903 (2001)
15. Ott, E., et al.: A local ensemble Kalman filter for atmospheric data assimilation. Tellus **56A**, 415–428 (2004)
16. Szunyogh, I., Kostelich, E.J., Gyarmati, G., Patil, D.J., Hunt, B.R., Kalnay, E., Ott, E., Yorke, J.A.: Assessing a local ensemble Kalman filter: perfect model experiments with the National Centers for Environmental Prediction global model. Tellus **57A**, 528–545 (2005)
17. Kalnay, E.: Atmospheric modelling: data assimilation and predictability. Cambridge University Press, Cambridge (2003)
18. Anderson, J.L., Anderson, S.L.: A Monte Carlo implementation of the nonlinear filtering problem to produce ensemble assimilations and forecasts. Mon. Weather. Rev. **126**, 2741–2758 (1999)
19. Oke, P.R., Sakov, P., Corney, S.P.: Impacts of localisation in the EnKF and EnOI: experiments with a small model. Ocean. Dyn. **57**, 32–45 (2007)
20. Schur, I.: Bemerkungen zur theorie der beschrnkten bilinearformen mit unendlich vielen vernderlichen. Journal Fur Die Reine Und Angewandte Mathematik **140**, 1–28 (1911)
21. Horn, R.: The Hadamard product. In: Johnson, C.R. (ed.) Matrix theory and applications, American Mathematical Society, Proceedings of Symposia in Applied Mathematics, vol. 40, pp. 87–169 (1990)
22. Gaspari, G., Cohn, S.E.: Construction of correlation functions in two and three dimensions. Q. J. R. Meteorol. Soc. **125**, 723–757 (1999)

Chapter 5
Applications

Abstract Some times the applications are neglected, when, instead of they are the true benchmark of the theory. This chapter deals, by using simple pedagogical examples, with the forecasting processes based on data assimilation methods. They are applied in different fields: the atmospheric study by the classical Lorenz model, the biology of cells by a model of tumor growth, the planetary general circulation by an application on Mars and the earthquake forecasts by renewal processes models.

The Kalman Filter equations describe how information from forecasts and from observations should be combined in a optimal way which extracts maximum information from each source. A forecast starts from initial conditions at time t_0 and run to time t_1 accumulating errors over this period, In language of DA the assimilated forecast/observation is called analysis state. The analysis state at t_1 is then used as initial condition for another forecast to time t_2 which is then combined with the next batch of observations at that time. When one uses the KF one describes not only the analysis state but also the uncertainties correlated with the information provided by the observations.

The Kalman filter presents the optimal solution to the data assimilation problem under the assumptions of linear models with Gaussian observation and model noise. These assumptions are strongly violated in stochastic point process models for earthquake forecasting where one needs a more general approach based on propagating the entire probability distribution, rather than solely mean and covariance.

This chapter treats, by using simple pedagogical examples, the forecasting processes based on data assimilation methods in different fields: the atmospheric study by the classical Lorenz model, the biology of cells by a model of tumor growth, the planetary general circulation by an application on Mars and the earthquake forecasts by renewal processes models.

5.1 Lorenz Model

The first of the pedagogical model is the simplified one proposed by Lorenz. Lorenz [1] discusses three one-dimensional toy models that incorporate many features shown in real atmospheric dynamics and in global numerical weather prediction models.

R. Guzzi, *Data Assimilation: Mathematical Concepts and Instructive Examples*,
SpringerBriefs in Earth Sciences, DOI 10.1007/978-3-319-22410-7_5

The first model (Lorenz model 1) was originally introduced in Lorenz [2] and Lorenz and Emanuel [3]. This model has become the standard model for the initial testing of EnKF schemes. The popularity of the model is in part due to the similarity between the propagation of uncertainties (forecast errors) in Lorenz model 1 and global circulation models in the midlatitude storm-track regions. In particular, the errors are propagated by dispersive waves whose behavior is similar to that of synoptic- scale Rossby waves, and the magnitude of the errors has a doubling time of about 1.5 days (where the dimensionless model time has been converted to dimensional time by assuming that the characteristic dissipation time scale in the real atmosphere is 5 days). Lorenz model 2 adds the feature of a smooth spatial variation of the model variables that resembles the smooth variation of the geopotential height streamfunction at the synoptic and large scales in the atmosphere. Lorenz model 3, the most refined and realistic of the three models in Lorenz [1], adds a rapidly varying small-amplitude component to the smooth large-scale flow, mimicking the effects of small-scale atmospheric processes.

Previously, Edward Lorenz [4] had described the convection motion of a fluid in a small, idealized Rayleigh-Benard cell. The Lorenz equations were derived from the Oberbeck-Boussinesq approximation to the equations describing fluid circulation in a shallow layer of fluid, heated uniformly from below and cooled uniformly from above. This fluid circulation is known as Rayleigh-Bénard convection. The fluid is assumed to circulate in two dimensions (vertical and horizontal) with periodic rectangular boundary conditions.

The partial differential equations modeling the system's stream function and temperature are subjected to a spectral Galerkin approximation: the hydrodynamic fields are expanded in Fourier series, which are then severely truncated to a single term for the stream function and two terms for the temperature. This reduces the model equations to a set of three coupled, nonlinear ordinary differential equations. A detailed derivation may be found, for example, in nonlinear dynamics textbooks. The Lorenz system is a reduced version of a larger system studied earlier by Barry Saltzman [5]. In his idealized model the boundary conditions of the fluid at the upper and lower plates were "stress free" rather than the realistic "no-slip", while the lateral boundary conditions are taken to be "periodic" rather than corresponding to realistic side walls, and the motion is assumed to be two-rather than three- dimensional. These modifications greatly simplified the mathematical analysis because the governing equations reduce the complicated partial differential equations describing the fluid motion and heat flow to three ordinary differential equations where the fluid and heat equations are coupled. Since these equations are non-linear, there will be terms coupling the different modes, and also terms generating higher harmonics representing the thermal modes and the fluid components velocity. The major approximation is that the latter terms are ignored.

$$\frac{dx}{dt} = f_x(x, y, z, \sigma) = -\sigma(x - y)$$
$$\frac{dy}{dt} = g_x(x, y, z, \rho) = \rho x - y - xz$$
$$\frac{dz}{dt} = h_x(x, y, z, \beta) = xy - \beta z, \tag{5.1}$$

where σ depends on the properties of the fluid (the ratio of the viscous to thermal diffusivities) and $\beta = \frac{8}{3}$ (this would be different for a different choice of horizontal wavelength or roll diameter). The temperature difference ρ is the important control parameter: for $\rho < 1$ the solution at long times is asymptotic to $x = y = z = 0$, i.e. no convection. For $\rho > 1$ various chaotic solutions occur. Lorenz realized the importance of the aperiodic motion. This Lorenz model has been widely used for exploring many real world problems.

In the last few years an increasing number of physicists have directed their research activities toward investigating the complex behavior of nonlinear model systems described by seemingly simple deterministic differential or difference equations. In fact there is a wealth of similar systems of nonlinear differential equations and difference equations whose chaotic state is characterized by apparently random-looking motion on attractors displaying a complicated structure. The trajectories in the phase space of these systems and the bifurcation sequence leading to the erratic motion have been studied intensively by computer calculations. Lyapunov characteristic exponents ($\sim 1 - 15$), Poincarè maps, symbolic transition dynamics, and other mathematical tools have been used. Lyapunov exponents [6] measure the growth rates of generic perturbations, in a regime where their evolution is ruled by linear equations. Possible universal properties of the bifurcation sequences have been studied using scaling approaches. In mathematics the Lyapunov exponent or Lyapunov characteristic exponent of a dynamical system is a quantity that characterizes the rate of separation of infinitesimally close trajectories. Quantitatively, two trajectories in phase space with initial separation $\delta\mathbf{Z}_0$ diverge (provided that the divergence can be treated within the linearized approximation) at a rate given by $|\delta\mathbf{Z}(t)| \approx e^{\lambda t}|\delta\mathbf{Z}_0|$ where λ is the Lyapunov exponent.

The rate of separation can be different for different orientations of initial separation vector. Thus, there is a spectrum of Lyapunov exponents equal in number to the dimensionality of the phase space. It is common to refer to the largest one as the Maximal Lyapunov exponent (MLE), because it determines a notion of predictability for a dynamical system. A positive MLE is usually taken as an indication that the system is chaotic (provided some other conditions are met, e.g., phase space compactness). Note that an arbitrary initial separation vector will typically contain some component in the direction associated with the MLE, and because of the exponential growth rate, the effect of the other exponents will be obliterated over time.

Since the equations for the dynamics of x, y, z are first order and autonomous, one may consider this set of variables the "phase space". The dynamics at each point, specified by the "velocity in phase space" vector is unique. The evolution in time then traces out a path in the three dimensions. An immediate result is that phase space trajectories cannot cross.

The sensitive dependence on initial conditions found by Lorenz is now known affectionately as Lorenz's "butterfly effect". In fact in a later paper Lorenz remarked:

One meteorologist remarked that if the theory were correct, one flap of the sea gull's wings would be enough to alter the course of the weather forever.

By the time of Lorenz's talk at the December 1972 meeting of the American Association for the Advancement of Science in Washington, D.C. the sea gull had evolved into the more poetic butterfly—the title of his talk was: *Predictability: does the Flap of a Butterfly's wings in Brazil set off a Tornado in Texas?*

Lorenz's work was largely ignored for ten years, but can now be seen as a prescient beginning to the study of chaos. Using this model we have identified some characteristics of chaos, that we will want to quantify further: apparent randomness in the time variation, broad band components to the power spectrum, sensitive dependence on initial conditions.

It is now known that the Lorenz equations are not an accurate description of the original (idealized) convection system for temperature differences, expressed by the dimensionless measure ρ, large enough to yield chaos.

McLauglin and Martin [7] also showed that chaos is obtained for a three dimensional version, but of course the mode equations are then not simply the Lorenz equations.

An alternative approach has been to construct experimental systems for which the Lorenz equations are a good description. Since the Lorenz equations break down due to the excitation of higher spatial harmonics, one scheme has been used to investigate convection in a circular glass tube held vertically and heated over the lower half and cooled over the upper half. Another experimental system described by the Lorenz equations is the Rititake dynamo, i.e. a homopolar generator with the output fed back through inductors and resistors to the coil generating the magnetic field. The disc is driven by a constant torque. The coupled circuit and rotation equations can be reduced to the Lorenz form, and experiments indeed show an apparently chaotic reversal of the coil current and the magnetic field. Analogies with the earth's magnetic field, which shows irregular reversals on a time scale of millions of years, are certainly intriguing even though Glatzmaier and Roberts [8] suggest that a three mode truncation will not be a good approximation for the turbulent dynamics of the earth's interior.

5.1.1 Solution of Lorenz 63 Model

The Lorenz model can be solved using a discretized four order Runge-Kutta method [9]. Starting from our Eq. (5.1) we have:

$$
\begin{aligned}
x_{k+1} &= x_k + \Delta t(f_1 + 2f_2 + 2f_3 + f_4)\\
y_{k+1} &= y_k + \Delta t(g_1 + 2g_2 + 2g_3 + g_4)\\
z_{k+1} &= z_k + \Delta t(h_1 + 2h_2 + 2h_3 + h_4),
\end{aligned}
\tag{5.2}
$$

where

$$
\begin{aligned}
f_1 &= f_x(x, y, z, \sigma)\\
g_1 &= f_y(x, y, z, \rho)\\
h_1 &= f_y(x, y, z, \beta)\\
f_2 &= f_x(x + \Delta t\frac{f_1}{2}, y + \Delta t\frac{g_1}{2}, z + \Delta t\frac{h_1}{2}, \sigma)\\
g_2 &= f_y(x + \Delta t\frac{f_1}{2}, y + \Delta t\frac{g_1}{2}, z + \Delta t\frac{h_1}{2}, \rho)\\
h_2 &= f_z(x + \Delta t\frac{f_1}{2}, y + \Delta t\frac{g_1}{2}, z + \Delta t\frac{h_1}{2}, \beta)\\
f_3 &= f_x(x + \Delta t\frac{f_2}{2}, y + \Delta t\frac{g_2}{2}, z + \Delta t\frac{h_2}{2}, \sigma)\\
g_3 &= f_y(x + \Delta t\frac{f_2}{2}, y + \Delta t\frac{g_2}{2}, z + \Delta t\frac{h_2}{2}, \rho)\\
h_3 &= f_z(x + \Delta t\frac{f_2}{2}, y + \Delta t\frac{g_2}{2}, z + \Delta t\frac{h_2}{2}, \beta)\\
f_4 &= f_x(x + \Delta t\frac{f_3}{2}, y + \Delta t\frac{g_3}{2}, z + \Delta t\frac{h_3}{2}, \sigma)\\
g_4 &= f_y(x + \Delta t\frac{f_3}{2}, y + \Delta t\frac{g_3}{2}, z + \Delta t\frac{h_3}{2}, \rho)\\
h_4 &= f_z(x + \Delta t\frac{f_3}{2}, y + \Delta t\frac{g_3}{2}, z + \Delta t\frac{h_3}{2}, \beta),
\end{aligned}
\tag{5.3}
$$

where Δt is the model time step and k is the time step index.

Since the discretization introduces an error, it is usual to add to (5.2) a term of the form $\sqrt{\Delta t}\eta$ where $\eta = (\eta_x, \eta_y, \eta_z)^T \sim N(\mathbf{0}, \mathbf{Q})$ is assumed to be a normally distributed random vector with zero mean and error covariance $\mathbf{Q}$.

5.1.2 Lorenz Model and Data Assimilation

On the web there are some implementations of the Lorenz experiment. They are written in MATLAB (Kuhl and Kostelich [10]) for the LETKF method or in C++ (Bannister [11]) or IDL (Migliorini [12]) for the EnSRKF method. The reason is that Lorenz 63 or 96, depending on the paper used, are a test bench to understand the behavior of a simplified atmosphere or to make some experiments of data assimilation or chaos.

In this paragraph we show the EnSRKF approach by Bannister. His approach is a little bit different from the general approach reported in the previous chapter, but it has the advantage to show another solution in the framework of what we have called the Deterministic Ensemble Kalman Filter.

At time t the forecast state of the system is represented by an n-element state vector $\mathbf{x}_k^f(t)$. Suppose the ensemble of N such model states exist ($1 \leq k \leq N$) and that the state vectors comprise the columns of the matrix $\mathbf{X}^f$ $(n \times N)$ as $\mathbf{X}^f(t) = \{\mathbf{x}_1^f, \mathbf{x}_2^f, \ldots, \mathbf{x}_N^f\}$ where the forecast ensemble will represent the realizations of the real system. The uncertainty of the forecasts is represented by the spread of the members.

Let's suppose the observations are available at time t in the p-element vector $\mathbf{y}(t)$, for p- observations.

The EnSRKF may be used to update the members to reduce the spread of the ensemble consistent with the uncertainties of the forecasts and the observations.

The Kalman filter is

$$\mathbf{x}_k^a = \mathbf{x}_k^f + \mathbf{P}^f\mathbf{H}^T(\mathbf{H}\mathbf{P}^f\mathbf{H}^T + \mathbf{R})^{-1}(\mathbf{y} - \mathbf{H}\mathbf{x}_k^f). \tag{5.4}$$

The EnSRKF process treats the whole ensemble as a single entity so the value of $\mathbf{x}^a$ is in fact the same of the ensemble mean. In matrix notation we have $\mathbf{X}^f$ and also the analysis matrix $\mathbf{X}^a$. The mean of $\mathbf{X}^f$ is denoted with $\bar{\mathbf{X}}^f$, of size $n \times N$, that is made up of N identical columns each containing the ensemble mean state vectors for model forecast. This matrix is that one of writing $\bar{\mathbf{x}}^f$. $\bar{\mathbf{X}}^a$ of size $n \times N$ is the ensemble mean matrix of the analyzed Kalman filter output and contains the writing of $\bar{\mathbf{x}}^a$.

In order to use the matrix notation we define also a new matrix $\mathbf{Y}$ that is made up of N identical columns each containing the observation vector $\mathbf{y}$. Thus

$$\bar{\mathbf{X}}^a = \bar{\mathbf{X}}^f + \mathbf{P}^f\mathbf{H}^T[\mathbf{H}\mathbf{P}^f\mathbf{H}^T + \mathbf{R}]^{-1}(\mathbf{Y} - \mathbf{H}\bar{\mathbf{X}}^f). \tag{5.5}$$

First of all we need to define $\mathbf{P}^f$ that is unknown. However it can be approximated because it is derived from statistical ensemble.

$$\begin{aligned}
\mathbf{P}^f &\approx \frac{1}{N-1}\sum_{k=1}^{N}(\mathbf{x}_k^f - \bar{\mathbf{x}}^f)(\mathbf{x}_k^f - \bar{\mathbf{x}}^f)^T \\
\mathbf{P}^f &\approx \frac{1}{N-1}\hat{\mathbf{X}}^f(\hat{\mathbf{X}}^f)^T,
\end{aligned} \tag{5.6}$$

where $\hat{\mathbf{X}}^f = (\mathbf{x}_i^f - \bar{\mathbf{x}}^f)$, $\quad k = 1, 2, 3, \ldots, N$ is the $n \times N$ matrix of the ensemble member perturbations. The same can be done for the analysis error covariance matrix:

$$\begin{aligned}
\mathbf{P}^a &\approx \frac{1}{N-1}\sum_{k=1}^{N}(\mathbf{x}_k^a - \bar{\mathbf{x}}^a)(\mathbf{x}_k^a - \bar{\mathbf{x}}^a)^T \\
\mathbf{P}^a &\approx \frac{1}{N-1}\hat{\mathbf{X}}^a(\hat{\mathbf{X}}^a)^T.
\end{aligned} \tag{5.7}$$

Substituting in (5.5) we obtain:

$$\bar{\mathbf{X}}^a = \bar{\mathbf{X}}^f + \frac{\hat{\mathbf{X}}^f(\hat{\mathbf{X}}^f)^T}{N-1}\mathbf{H}^T[\mathbf{H}\frac{\hat{\mathbf{X}}^f(\hat{\mathbf{X}}^f)^T}{N-1}\mathbf{H}^T + \mathbf{R}]^{-1}(\mathbf{Y} - \mathbf{H}\bar{\mathbf{X}}^f). \tag{5.8}$$

Rearranging and simplifying we obtain:

$$\bar{\mathbf{X}}^a = \bar{\mathbf{X}}^f + \hat{\mathbf{X}}^f(\hat{\mathbf{X}}^f)^T\mathbf{H}^T[\mathbf{H}\hat{\mathbf{X}}^f(\hat{\mathbf{X}}^f)^T\mathbf{H}^T + (N-1)\mathbf{R}]^{-1}(\mathbf{Y} - \mathbf{H}\bar{\mathbf{X}}^f). \tag{5.9}$$

Indicating $\mathbf{B} = \mathbf{H}\hat{\mathbf{X}}^f$ we have:

$$\bar{\mathbf{X}}^a = \bar{\mathbf{X}}^f + \hat{\mathbf{X}}^f\mathbf{B}^T[\mathbf{B}\mathbf{B}^T + (N-1)\mathbf{R}]^{-1}(\mathbf{Y} - \mathbf{H}\bar{\mathbf{X}}^f). \tag{5.10}$$

If we indicate with $\mathbf{C} = (\mathbf{B}\mathbf{B}^T + (N-1)\mathbf{R})$ we can write:

$$\bar{\mathbf{X}}^a = \bar{\mathbf{X}}^f + \hat{\mathbf{X}}^f\mathbf{B}^T\mathbf{C}^{-1}(\mathbf{Y} - \mathbf{H}\bar{\mathbf{X}}^f). \tag{5.11}$$

Now we turn to the state covariance. Remembering that $\mathbf{P}^a = (\mathbf{I} - \mathbf{K}\mathbf{H})\mathbf{P}^f = \mathbf{P}^f - \mathbf{K}\mathbf{H}\mathbf{P}^f$, substituting in it for $\mathbf{K} = \mathbf{P}^f\mathbf{H}^T[\mathbf{H}\mathbf{P}^f\mathbf{H}^T + \mathbf{R}]^{-1}$, we get:

$$\mathbf{P}^a = \mathbf{P}^f - \mathbf{P}^f\mathbf{H}^T[\mathbf{H}\mathbf{P}^f\mathbf{H}^T + \mathbf{R}]^{-1}\mathbf{H}\mathbf{P}^f. \tag{5.12}$$

Remembering that $\mathbf{P}^f$ and $\mathbf{P}^a$ are given from (5.6) and (5.7), and taking into account that $\mathbf{H}\hat{\mathbf{X}}^f = \mathbf{B}$ we make a substitution in the relation (5.12) obtaining:

$$\hat{\mathbf{X}}^a(\hat{\mathbf{X}}^a)^T = \hat{\mathbf{X}}^f(\hat{\mathbf{X}}^f)^T - \hat{\mathbf{X}}^f\mathbf{B}^T[\mathbf{B}\mathbf{B}^T + \mathbf{R}(N-1)]^{-1}\mathbf{B}(\hat{\mathbf{X}}^f)^T. \tag{5.13}$$

Introducing a new term

$$\mathbf{G} = \mathbf{B}^T[\mathbf{B}\mathbf{B}^T + \mathbf{R}(N-1)]^{-1}\mathbf{B} \tag{5.14}$$

we obtain

$$\hat{\mathbf{X}}^a(\hat{\mathbf{X}}^a)^T = \hat{\mathbf{X}}^f[\mathbf{I} - \mathbf{G}](\hat{\mathbf{X}}^f)^T. \tag{5.15}$$

In the square root scheme the idea is to obtain an ensemble perturbed analysis (contained into $\hat{\mathbf{X}}^a$) that has the covariance given by the relation (5.12). The matrix $\hat{\mathbf{X}}^a$ is thought as the "square root" of the analysis error covariance matrix given by (5.7) that can be added to the mean $\bar{\mathbf{X}}^a$ of relation (5.10). The last step is to find $\hat{\mathbf{X}}^a$ that has the properties of (5.7), which is the same as to find the square root of (5.7). Thus we decompose firstly $\mathbf{G} = \mathbf{\Lambda}\mathbf{\Sigma}C$ in its eigenvectors $\mathbf{\Sigma}$ and eigenvalues $\mathbf{\Lambda}$.

$$\hat{\mathbf{X}}^a(\hat{\mathbf{X}}^a)^T = \hat{\mathbf{X}}^f[\mathbf{I} - \mathbf{\Lambda}\mathbf{\Sigma}\mathbf{\Lambda}^T](\hat{\mathbf{X}}^f)^T \tag{5.16}$$

Bringing the eigenvectors outside brackets and using the properties of eigenvectors $\mathbf{\Lambda I \Lambda}^T = \mathbf{I}$ we have:

$$\hat{\mathbf{X}}^a(\hat{\mathbf{X}}^a)^T = \hat{\mathbf{X}}^f \mathbf{\Lambda}[\mathbf{I} - \mathbf{\Sigma}]\mathbf{\Lambda}^T(\hat{\mathbf{X}}^f)^T \tag{5.17}$$

The square matrix $[\mathbf{I} - \mathbf{\Sigma}]$ has an infinite number of possible roots. One is simply $[\mathbf{I} - \mathbf{\Sigma}]^{\frac{1}{2}}$ and the other is $[\mathbf{I} - \mathbf{\Sigma}]^{\frac{1}{2}}\mathbf{\Lambda}^T$ since $\mathbf{\Lambda}^T\mathbf{\Lambda} = \mathbf{I}$. Therefore:

$$[(\mathbf{I} - \mathbf{\Sigma})^{\frac{1}{2}}\mathbf{\Lambda}^T][(\mathbf{I} - \mathbf{\Sigma})^{\frac{1}{2}}\mathbf{\Lambda}^T]^T = (\mathbf{I} - \mathbf{\Sigma})^{\frac{1}{2}}\mathbf{\Lambda}^T\mathbf{\Lambda}(\mathbf{I} - \mathbf{\Sigma})^{\frac{1}{2}} = (\mathbf{I} - \mathbf{\Sigma})^{\frac{1}{2}}(\mathbf{I} - \mathbf{\Sigma})^{\frac{1}{2}} = \mathbf{I} - \mathbf{\Sigma} \tag{5.18}$$

Thus using the second square root we have:

$$\hat{\mathbf{X}}^a(\hat{\mathbf{X}}^a)^T = \hat{\mathbf{X}}^f \mathbf{\Lambda}[\mathbf{I} - \mathbf{\Sigma}]^{\frac{1}{2}}\mathbf{\Lambda}^T([\mathbf{I} - \mathbf{\Sigma}]^{\frac{1}{2}})^T\mathbf{\Lambda}^T(\hat{\mathbf{X}}^f)^T \tag{5.19}$$

from which we can extract $\hat{\mathbf{X}}^a$.

$$\hat{\mathbf{X}}^a = \hat{\mathbf{X}}^f \mathbf{\Lambda}[\mathbf{I} - \mathbf{\Sigma}]^{\frac{1}{2}}\mathbf{\Lambda}^T \tag{5.20}$$

The final step is to construct the full ensemble from the perturbation and then to propagate this ensemble to the next step to obtain the final next step.

$$\mathbf{X}^a(t) = \bar{\mathbf{X}}^a + \hat{\mathbf{X}}^a \tag{5.21}$$

$$\mathbf{X}^f(t + \Delta t) = \mathcal{M}(\mathbf{X}^a(t)) + \delta\mathbf{X}(t) \tag{5.22}$$

where $\mathcal{M}$ is the non linear forecast model and $\delta\mathbf{X}(t)$ is a $n \times N$ matrix of stochastic perturbations to simulate the imperfect model with specified error covariance $\mathbf{Q}$. In absence of observations no data assimilation can be performed and only the forecast is performed.

5.2 Biology and Medicine

In recent years, a branch of science called Systems Biology has developed. It is a discipline intended to redefine the concept of biological system in term of integrated observable entities and to analyze complex data sets using interdisciplinary technology platforms as: Phenomics, Genomics, Epigenomics, Transcriptomics, Interferomics, Proteomics, Metabolomics, etc. The model at the center of observation returns to be the cell and cell populations, together with its microenvironment. The approach of System Biology tends to promote a comprehensive vision that can combine complex causal relationships. Such relationships are not simply associated to the classical concept of deterministic causality, but they rather correspond to the concept of probabilistic causality, common in physics.

The System Biology considers that information flow depends on a mix of both molecular (DNA, protein, lipids, etc.) and physical signals (System status parameters: strength, thermodynamic constraints), which interact with each other through a non-linear dynamics. To understand these interactions is necessary, on one side, to provide data on which it is possible to conduct statistical evaluations, made possible through techniques of Genomics, Proteomics, Metabolomics, etc. Such approaches provide large amounts of information and the way to understand how physical forces, the weight of the tangential stress, stiffness, surface tension and gravitational interactions, determine the fate of complex biological systems.

At the same time mathematical and numerical models have been progressively used as a tool for supporting medical research in the biology and medicine science. *Silico* experiments have provided remarkable insights into a physio-pathological process completing the traditional in *vitro* and in *vivo* investigations. Furthermore numerical models have been used to give a dynamical representation of the biology of a specific patient and to be a support to the prognostic activity.

The need to obtain not only qualitative responses but also quantitative responses for diagnostic purposes has stimulated the design of new methods and instruments of measurements and imaging. The advent of high resolution 3D imaging instruments in biology and medicine are producing novel approaches to treat huge amount of data which can be used for the numerical simulations. However, beyond the validation, it is possible to merge simulations and measures by means of more sophisticated numerical techniques.

Recently some Data Assimilation methods have been applied to biomathematics, merging observed (generally sparse and noisy) information into a numerical model. This approach improves the quality of the information because allows to include effects otherwise difficult to modeling introducing a sophisticated filter able to balance the uncertainties of the measured data with basic principles. In summary, data assimilation methods born in geophysics and meteorology, are also mature to be used in the fields of biology and medicine.

What is relevant is the transparency of the mathematical and statistical methodology and of the digital representation of a specific pathology on that patient (electronic/virtual phenotype). This approach allows us to initiate a route towards a well based selection of therapeutics for the future integrated and personalized care in agreement with the actual institution guidelines development.

The integration of missed information such as those of case report will allow to approach the solution of a single difficult case by using a set of different domain choices (social, ethical, medical etc.) capable to improve quality of life with the specific pathology and patient stage.

The data from specific research in the pathology sector will be the base for the continuous and future tuning of the data base and for the periodic validation of a complex virtual engine. By reconstructing the electronic phenotype, the individuality recognition (personalization) of each single patient is actually linked to the overall information recruited on the disease.

Phenotype domains are crucial to see the personalized patient represented in all its own dimensions even though social economic psychological-ethical condition could be prone to orient the final selection of the integrated therapeutic actions.

5.2.1 Tumor Growth

Cancer may be regarded as a paradigmatic microcosmos for all of biology, i.e. as an observable system where mutation and evolution take place. This is a multiscale-problem in the life sciences.

A general approach to the problem of tumor growth is proposed by Chaplain [13]. Starting from a macroscopic model it is possible to describe the space-temporal growth and spread of the tumor by classical deterministic reaction—diffusion equations.

Consider the flux of cells across a volume V enclosed by a smooth surface S in term of mass conservation, the rate at which the number of cells changes within V is equal to the number of cells across S. Since the process is affected by creation of cells due to mitosis or a lost due to their death we need also to take into account these effects.

By Fick's second law one can predict how diffusion changes the concentration of cells with time.

$$\frac{\partial \mathbf{c}}{\partial t} = D\frac{\partial^2 \mathbf{c}}{\partial x^2}, \tag{5.23}$$

where $\mathbf{c}$ is the concentration $(amount\ of\ substance) \times length^{-3}$, $(\frac{\mathbf{mol}}{m^3})$; it depends on location x in (m) and time t in sec, D is the diffusion coefficient $m^2 \times time^{-1}$ Taking into account the process of creation and/or death of the cells due to treatment, one can write the partial differential equation in three dimension

$$\frac{\partial \mathbf{c}}{\partial t} = \frac{\partial}{\partial x}(D\frac{\partial \mathbf{c}}{\partial x}) + \frac{\partial}{\partial y}(D\frac{\partial \mathbf{c}}{\partial y}) + \frac{\partial}{\partial z}(D\frac{\partial \mathbf{c}}{\partial z}) + \alpha\mathbf{c}(1 - \frac{\mathbf{c}}{T_{max}}) - \mathbf{G}(t)\mathbf{c}, \tag{5.24}$$

where $\mathbf{G}(t)$ accounts for the temporal profile of treatment and as a first approximation may be considered a constant and α is the maximum glioma growth rate. Such model is given both by Chaplain [13] and Swanson [14]. In order to avoid the exponential growth term, the cells growth logistically with some carrying capacity T_{max} at any given point of the model's domain. The representative parameters for this model applied to brain are given in the following Table 5.1, where are reported also the density of white matter, gray matter, corpus callosum and cerebro spinal fluid (CSF). The initial conditions are

$$\mathbf{c}(x, y, x, 0) = \mathbf{f}(x, y, z) \tag{5.25}$$

while the boundary conditions on the brain domain on which the diffusion equation is to be solved is

Table 5.1 Parameters for Swanson model

Parameter	Meaning	Value
α	Maximum glioma growth rate day^{-1}	0.2
T_{max}	Glioma carrying capacity cells mm^{-2}	1000
D white matter	Diffusion rate (mm^{2}day^{-1})	0.0065
D corpus callosum	Diffusion rate (mm^{2}day^{-1})	0.001
D gray matter	Diffusion rate (mm^{2}day^{-1})	0.0013
D CSF	Diffusion rate (mm^{2}day^{-1})	0.001

$$\hat{\mathbf{n}} \cdot D(\frac{\partial \mathbf{c}}{\partial x}\hat{\mathbf{i}} + \frac{\partial \mathbf{c}}{\partial y}\hat{\mathbf{j}} + \frac{\partial \mathbf{c}}{\partial z}\hat{\mathbf{k}}) = 0 \quad on \quad \partial\Omega. \tag{5.26}$$

In the Neumann boundary condition the normal derivative of the unknown function **c** (solution of PDE) is

$$\frac{\partial \mathbf{c}}{\partial n}(x, y, z) = \hat{n} \cdot \nabla \mathbf{c} = \mathbf{f}(x, y, z) \quad \forall \in \partial\Omega, \tag{5.27}$$

where $\hat{\mathbf{n}}$ is the unit vector normal to the boundary $\partial\Omega$ of the domain Ω, $\hat{\mathbf{i}}, \hat{\mathbf{j}}, \hat{\mathbf{k}}$ are the vectors representing the subintervals in which the coordinates are subdivided and $\mathbf{f}(x, y, z)$ is a known function that defines the initial spatial distribution of malignant cells [15].

In order to solve the Eq. (5.24) one uses the Crank-Nicolson method, well known in the frame of many diffusion problems.

5.2.2 Growth Tumor Data Assimilation with LETKF

We report here the paradigmatic approach of Kostelich et al. [16], because the data assimilation procedure, despite being applied to a brain tumor (Glioblastoma multiform-GBM-), does not depend on the details of a given cancer growth model and should be applicable to other models of cancer or biological phenomena. They report two types of models. The one reported here is simpler than the other which involves more processes, from haptotaxis to degradation and repair.

We prefer to avoid discussing the second model in order to emphasize the data assimilation methods. The equation has been integrated on the brain geometry obtained from the Brain web data base developed by the McConnel Brain Imaging Center of the Montreal neurological Institute at McGill University.

This is an example of Local Ensemble Transform Kalman Filter (see Hunt et al. [17]) one of the general approach of classical or advanced data assimilation used in weather forecast. It can be addressed to GBM's growth even thought the details of the growth of the tumor cells are poorly known due to complexity of the mechanisms involved. Data are obtained from high resolution episodic images that are obtained at intervals of weeks to months using chemical agents to enhance the contrast. The therapy induces further complications that affect the information we need to apply the data assimilation methods. However the goal of the research is to obtain good quantitative predictions of GBM growth and spread as well as to estimate their uncertainties.

Data assimilation have been obtained observing the magnetic resonance images and using two different model of growth of the tumor.

Even thought the method is the one we have defined in EnKF we exploit and expand it here.

The problem is to estimate the solution trajectory that best fits the observations, given an imperfect forecast model that produces trajectories from time t_{n-1} to t_n.

Assuming at each time t_k, $k = 1, \ldots, n$ that the observation is related to the operator $\mathbf{H}$, i.e. $\mathbf{y}_k = \mathbf{H}(\mathbf{x}(t_k)) + \mathbf{v}$, where $\mathbf{v} = N(\mathbf{0}, \mathbf{R})$ is a Gaussian random vector, and that the system evolves according to a linear model $\mathbf{x}_n = \mathbf{M}_n \mathbf{x}_{n-1}$, the problem is to maximize the likelihood function

$$\mathbf{L}[\mathbf{x}(t)] = \prod_{j=1}^{n} \exp\{-\frac{1}{2}[\mathbf{y}_j - \mathbf{H}(\mathbf{x}(t_j))]^T \mathbf{R}^{-1} [\mathbf{y}_j - \mathbf{H}(\mathbf{x}(t_j))]\} \tag{5.28}$$

or minimize the cost function obtained taking the log of relation (5.28)

$$\mathbf{J}[\mathbf{x}(t)] = \sum_{j=1}^{n} \{[\mathbf{y}_j - \mathbf{H}(\mathbf{x}(t_j))]^T \mathbf{R}^{-1} [\mathbf{y}_j - \mathbf{H}(\mathbf{x}(t_j))]\}. \tag{5.29}$$

Since we have seen the extension of Kalman Filter to a non linear scenario implies that the propagation of the error analysis covariance is no longer traceable by the background covariance matrix given by $\mathbf{P}_n^f = \mathbf{M}_n \mathbf{P}_{n-1}^a (\mathbf{M}_n)^T$, one of the possible solutions is to select an ensemble of k possible trajectories whose variance is approximated by $\mathbf{P}_{n-1}^a$. Each ensemble is updated by the model to time t_n to compute the updated ensemble background/forecast covariance $\mathbf{P}_n^f$. Accurate analysis is obtained when the spread of the ensemble approximates well $\mathbf{P}_n^f$, otherwise the analysis fails to correct errors in forecast model. LETKF uses localization to overcome problems arising from the size of trajectories. Since the size of k is in general smaller than the model resolution m, the LEFKF's strategy is to perform the analysis at each point individually by forming a local ensemble over a subset of the model domain (Hunt et al. [17]). In this way the dynamics at a given point is captured over the local region.

Then the local ensemble will estimate the background uncertainty and will correct the forecast at each point, giving an updated global analysis over the entire model grid. The advantage of LETKF is to be model independent, i.e. the analysis is computed without the use of the model equations. This is a key feature when the system under study is poorly understood and the models are not fully mature.

The procedure starts at time t_{n-1} with the analysis of the ensemble consisting of m-dimensional model vectors

$$\{\mathbf{x}_{t_{n-1}}^{a_i} \quad i = 1, 2, \ldots, k\}, \tag{5.30}$$

that represents the density of the cells (proliferating, migrating, chemo-repelling) and the extracellular matrix (ECM) at each grid point of the model geometry assumed for the cavity in which the tumor is growing.

The mean is considered as the best value estimate of the most likely state of the system. The update of the ensemble member is abstained from the model up the time t_n. In such a way one can obtain the forecast ensemble

$$\{\mathbf{x}_{t_n}^{f_i} \quad i = 1, 2, \ldots, k\}. \tag{5.31}$$

The relation between the forecast state and the analysis state is given by:

$$\mathbf{x}_{t_n}^{f_i} = \mathbf{F}(\mathbf{x}_{t_{n-1}}^{a_i}, t_{n-1}) \quad i = 1, 2, \ldots, k. \tag{5.32}$$

From now on the time subscripts will be omitted. The forecast mean is:

$$\bar{\mathbf{x}}^f = k^{-1} \sum_{i=1}^{k} \mathbf{x}^{f_i}, \tag{5.33}$$

with the appropriate error covariance matrix

$$\begin{aligned} \mathbf{P}^f &= (k-1)^{-1} \sum_{i=1}^{k} (\mathbf{x}^{f_i} - \bar{\mathbf{x}}^f)(\mathbf{x}^{f_i} - \bar{\mathbf{x}}^f)^T \\ &= (k-1)^{-1} \mathbf{X}^f (\mathbf{X}^f)^T, \end{aligned} \tag{5.34}$$

where the $m \times k$ forecast ensemble perturbation matrix $\mathbf{X}^f$ has the i-column $(\mathbf{x}^{f_i} - \bar{\mathbf{x}}^f)$.

The problem is to compute the analysis ensemble $\{\mathbf{x}^{a_i} \quad i = 1, 2, \ldots, k\}$ with appropriate mean and error covariance matrix

$$\bar{\mathbf{x}}^a = k^{-1} \sum_{i=1}^{k} \mathbf{x}^{a_i} \tag{5.35}$$

and

$$\begin{aligned} \mathbf{P}^f &= (k-1)^{-1} \sum_{i=1}^{k} (\mathbf{x}^{a_i} - \bar{\mathbf{x}}^a)(\mathbf{x}^{a_i} - \bar{\mathbf{x}}^a)^T \\ &= (k-1)^{-1} \mathbf{X}^a (\mathbf{X}^a)^T, \end{aligned} \tag{5.36}$$

since the task of LETFK is to minimize the cost function

$$\mathbf{J}[\mathbf{x}(t)] = [\mathbf{y} - \mathbf{H}(\mathbf{x})]^T \mathbf{R}^{-1} [\mathbf{y} - \mathbf{H}(\mathbf{x})] + (\mathbf{x} - \bar{\mathbf{x}}^f)^T (\mathbf{P}^f)^{-1} (\mathbf{x} - \bar{\mathbf{x}}^f). \tag{5.37}$$

However the rank of the covariance forecast matrix (5.34) can be at most $k - 1$ and therefore it is not invertible. Nevertheless its inverse is well defined on the space S defined by the forecast ensemble perturbation, that is the column of $\mathbf{X}^f$. Then $\mathbf{J}$ is well defined for $(\mathbf{x} - \bar{\mathbf{x}}^f)$ in S and the minimization can be carried out in this subspace. Thus in order to carry out the analysis on S one needs to choose an appropriate coordinate system through an affine transformation. If one regards $\mathbf{X}^f$ as a linear transform from a k-dimensional space $\tilde{S}$ onto S, one may perform the analysis in $\tilde{S}$. Then let $\mathbf{w}$ denote a vector in the $\tilde{S}$ space. $\mathbf{X}^f\mathbf{w}$ belongs to the space S spanned by the forecast perturbation ensemble and $\mathbf{x} = \bar{\mathbf{x}}^f + \mathbf{X}^f\mathbf{w}$ is the corresponding model status. Note that if $\mathbf{w}$ is a random Gaussian vector with mean $\mathbf{0}$ and covariance $(k-1)^{-1}\mathbf{I}$ then $\mathbf{x} = \bar{\mathbf{x}}^f + \mathbf{X}^f\mathbf{w}$ is Gaussian with mean $\bar{\mathbf{x}}$ and covariance $\mathbf{P}^f = (k-1)^{-1}\mathbf{X}^f(\mathbf{X}^f)^T$.

Then the cost function is on $\tilde{S}$:

$$\tilde{\mathbf{J}}(\mathbf{w}) = [\mathbf{y} - \mathbf{H}(\bar{\mathbf{x}}^f - \mathbf{X}^f\mathbf{w})]^T \mathbf{R}^{-1} [\mathbf{y} - \mathbf{H}(\bar{\mathbf{x}}^f - \mathbf{X}^f\mathbf{w})] + (k-1)\mathbf{w}^T\mathbf{w}. \tag{5.38}$$

If $\bar{\mathbf{w}}^a$ minimizes $\tilde{\mathbf{J}}$ then $\bar{\mathbf{x}}^a = \bar{\mathbf{x}}^f + \mathbf{X}^f\mathbf{w}^a$ minimizes $\mathbf{J}$. In order to prove this, let $\mathbf{P}$ be the orthogonal projection matrix from $\tilde{S}$ onto the subspace spanned by the columns of $\mathbf{X}^f$, that is $\mathbf{P} = \mathbf{X}^f[(\mathbf{X}^f)^T\mathbf{X}^f]^{-1}(\mathbf{X}^f)^T$. Decompose the vector $\mathbf{w}$ as $\mathbf{w} = \mathbf{P}\mathbf{w} + (\mathbf{I} - \mathbf{P})\mathbf{w}$. Substituting into the first term of the cost function (5.38) one obtains

$$\begin{aligned}(k-1)\mathbf{w}^T\mathbf{w} &= (k-1)\mathbf{w}^T[\mathbf{P}\mathbf{w} + (\mathbf{I} - \mathbf{P})\mathbf{w}] \\ &= (k-1)\mathbf{w}^T\mathbf{P}\mathbf{w} + (k-1)^{-1}\mathbf{w}^T(\mathbf{I} - \mathbf{P})\mathbf{w}.\end{aligned} \tag{5.39}$$

Since $\mathbf{w} = (\mathbf{X}^f)^{-1}(\mathbf{x} - \bar{\mathbf{x}}^f)$ and given the orthogonalization of $\mathbf{P}$ one gets:

$$\begin{aligned}(k-1)\mathbf{w}^T\mathbf{P}\mathbf{w} &= (k-1)[(\mathbf{X}^f)^{-1}(\mathbf{x} - \bar{\mathbf{x}}^f)]^T\mathbf{X}^f[(\mathbf{X}^f)^T\mathbf{X}^f]^{-1}(\mathbf{X}^f)^T(\mathbf{X}^f)^{-1}(\mathbf{x} - \bar{\mathbf{x}}^f) \\ &= (k-1)[(\mathbf{X}^f)^{-1}(\mathbf{x} - \bar{\mathbf{x}}^f)]^T(\mathbf{X}^f)^{-1}(\mathbf{x} - \bar{\mathbf{x}}^f) \\ &= (k-1)[(\mathbf{x} - \bar{\mathbf{x}}^f)]^T[(\mathbf{X}^f)^{-1}]^T(\mathbf{X}^f)^{-1}[(\mathbf{x} - \bar{\mathbf{x}}^f)] \\ &= (k-1)[(\mathbf{x} - \bar{\mathbf{x}}^f)]^T[\mathbf{X}^f(\mathbf{X}^f)^{-1}]^T[(\mathbf{x} - \bar{\mathbf{x}}^f)] \\ &= [(\mathbf{x} - \bar{\mathbf{x}}^f)]^T(\mathbf{P}^f)^{-1}[(\mathbf{x} - \bar{\mathbf{x}}^f)].\end{aligned} \tag{5.40}$$

Combining (5.39) and (5.40) into the cost function (5.38) one gets

$$\tilde{\mathbf{J}}(\mathbf{w}) = (k-1)\mathbf{w}^T(\mathbf{I} - \mathbf{P})\mathbf{w} + \mathbf{J}(\bar{\mathbf{x}}^f + \mathbf{X}^f\mathbf{w}), \tag{5.41}$$

where the first term on the right is the orthogonal projection of $\mathbf{w}$ onto the null space N of $\mathbf{X}^f$ which depends on the $\mathbf{w}$ components in the null space. The second term only depends on the components in the column space S, of $\mathbf{X}^f$ Then it follows that $\bar{\mathbf{w}}^a$ minimizes $\tilde{\mathbf{J}}$ if and only if it is orthogonal to N and $\bar{\mathbf{x}}^a$ minimizes $\mathbf{J}$

In order to derive the updated analysis and error covariance matrix one proceeds to compute a minimizer to $\tilde{\mathbf{J}}$ based on Kalman filter. First of all one linearizes the observation operator $\mathbf{H}$ to each forecast trajectory $\mathbf{x}^{f_i}$ producing the l dimensional vectors ($l \leq m$), representing the spatial dimension of the observations, which comprise the forecast observation ensemble,

$$\mathbf{y}^{f_i} = \mathbf{H}(\mathbf{x}^{f_i}). \tag{5.42}$$

As before let's denote the mean forecast observation with $\bar{\mathbf{y}}^f$ and the $l \times k$ forecast observation ensemble perturbation matrix with $\mathbf{Y}^f$ whose ith column is $\mathbf{x}^{f_i} - \bar{\mathbf{x}}^f$.

$$\mathbf{H}(\bar{\mathbf{x}}^f + \mathbf{X}^f \mathbf{w}) \approx \bar{\mathbf{y}}^f + \mathbf{Y}^f \mathbf{w}, \tag{5.43}$$

remembering the Kalman filter update equation we obtain:

$$\bar{\mathbf{w}} = \mathbf{K}[\mathbf{y} - \bar{\mathbf{y}}^f - \mathbf{Y}^f \mathbf{w}] \tag{5.44}$$

Remembering

$$\mathbf{K} = \mathbf{P}^a \mathbf{H}^T \mathbf{R}^{-1} \tag{5.45}$$

and

$$\mathbf{P}^a = (\mathbf{I} + \mathbf{P}^f \mathbf{H}^T \mathbf{R}^{-1} \mathbf{H})^{-1} \mathbf{P}^f, \tag{5.46}$$

since $\mathbf{Y}^f$ is playing the role of $\mathbf{H}$ one has:

$$\bar{\mathbf{w}}^a = \tilde{\mathbf{P}}^a (\mathbf{Y}^f)^T \mathbf{R}^{-1} [\mathbf{y} - \bar{\mathbf{y}}^f] \tag{5.47}$$

and

$$\tilde{\mathbf{P}}^a = [(k-1)\mathbf{I} + (\mathbf{Y}^f)^T \mathbf{R}^{-1} \mathbf{Y}^f]^{-1} \tag{5.48}$$

The analysis mean and error covariance matrix in model variables are:

$$\bar{\mathbf{x}}^a = \bar{\mathbf{x}}^f + \mathbf{X}^f \bar{\mathbf{w}}^a, \tag{5.49}$$

$$\mathbf{P}^a = \mathbf{X}^f \tilde{\mathbf{P}}^a (\mathbf{X}^f)^T. \tag{5.50}$$

Let's now determine an analysis ensemble whose mean and error covariance matrix are given from the previous last Eqs. (5.49) and (5.50). The strategic approach is based on the two following points.

1. Since LETKF adopts the localization, the ensemble depends continuously on the analysis covariance matrix, that ensures the analysis ensemble covariance matrix $\tilde{\mathbf{P}}^a$ is the same in any points of the grid. A possible choice is $\mathbf{X}^a = \mathbf{X}^f \mathbf{W}^a$ where

$\mathbf{W}^a$ is the symmetric square root matrix defined as $(k-1)\tilde{\mathbf{P}}^a = \mathbf{W}^a(\mathbf{W}^a)^T$. In fact the (5.50) is, taking into account also $\mathbf{X}^a = \mathbf{X}^f\mathbf{W}^a$

$$\begin{aligned}\mathbf{P}^a &= \mathbf{X}^f\tilde{\mathbf{P}}^a(\mathbf{X}^f)^T \\ &= \mathbf{X}^f(k-1)^{-1}\mathbf{W}^a(\mathbf{W}^a)^T(\mathbf{X}^f)^T \\ &= (k-1)^{-1}\mathbf{X}^a(\mathbf{X}^a)^T. \end{aligned} \tag{5.51}$$

2. Choose a matrix whose columns sum to zero and has the desired error covariance matrix. In order to show the columns of $\mathbf{X}^a$ sum to zero is equivalent to show $\mathbf{X}^a\mathbf{w} = \mathbf{0}$ where $\mathbf{w} \equiv (\mathbf{1}, \mathbf{1}, \ldots)^T$. Then because also $\mathbf{Y}^f$ sum to zero we have

$$\begin{aligned}(\tilde{\mathbf{P}}^a)^{-1}\mathbf{w} &= (k-1)\mathbf{w} + (\mathbf{Y}^f)^{-1}\mathbf{R}^{-1}\mathbf{Y}^f\mathbf{w} \\ &= (k-1)\mathbf{w}. \end{aligned} \tag{5.52}$$

Thus $\mathbf{w}$ is an eigenvector for $\tilde{\mathbf{P}}^a$ with eigenvalue $(k-1)^{-1}$. At the same time, from $\mathbf{P}^a\mathbf{w} = (k-1)^{-1}\mathbf{W}^a(\mathbf{W}^a)^T\mathbf{w}$, we have that $\mathbf{w}$ is also an eigenvector for $\mathbf{W}^a$ with eigenvalue $\mathbf{1}$. Since the columns of $\mathbf{X}^f$ sum to zero

$$\begin{aligned}\mathbf{X}^a\mathbf{w} &= \mathbf{X}^f\mathbf{W}^a\mathbf{w} \\ &= \mathbf{X}^f\mathbf{w} = \mathbf{0} \end{aligned} \tag{5.53}$$

3. Shift the columns by $\bar{\mathbf{x}}$ because the mean as in (5.49) holds. Add $\bar{\mathbf{w}}^a$ to each column vector $\mathbf{w}^{a_i}$ of $\mathbf{W}^a$ defined as $\tilde{\mathbf{x}}^{a_i}$. By these weighting vectors one gets the ith analysis ensemble in space model $\mathbf{x}^{a_i} = \bar{\mathbf{x}}^f + \mathbf{X}^a\tilde{\mathbf{x}}^{a_i}$.

With the previous positions, the updated analysis mean is

$$(k)^{-1}\sum_{i=1}^{k}\mathbf{x}^{a_i} = (k)^{-1}\sum_{i=1}^{k}(\bar{\mathbf{x}}^f + \mathbf{X}^a\tilde{\mathbf{x}}^{a_i})$$

$$\begin{aligned}\bar{\mathbf{x}}^a &= \bar{\mathbf{x}}^f + \mathbf{X}^f\bar{\mathbf{w}}^a + k^{-1}\mathbf{X}^f\sum_{i=1}^{k}\mathbf{w}^{a_i} \\ &= \bar{\mathbf{x}}^f + \mathbf{X}^f\bar{\mathbf{w}}^a + k^{-1}\mathbf{X}^f\mathbf{W}^a\mathbf{w} \\ &= \bar{\mathbf{x}}^f + \mathbf{X}^f\bar{\mathbf{w}}^a. \end{aligned} \tag{5.54}$$

5.2.3 LETFK Receipt Computation

Following Hunt [17] the procedure starts with several preliminary computation carried out over the entire model grid.

1. The observation operator is applied to the m dimensional forecast ensemble vector $\mathbf{x}^{f_i}$ to form the forecast observation ensemble $\mathbf{y}^{f_i}$
2. After both ensembles are averaged the vectors $\mathbf{y}^{f_i} - \bar{\mathbf{y}}^f$ and $\mathbf{x}^{f_i} - \bar{\mathbf{x}}^f$ that are used to form the perturbed matrix $\mathbf{Y}^f$ and $\mathbf{X}^f$ are computed.
3. For each local region select the components of $\mathbf{x}^{f_i}$, $\mathbf{y}^{f_i}$, $\mathbf{X}^f$, $\mathbf{Y}^f$. and $\mathbf{R}$.
4. Compute the $k \times l$ matrix $\mathbf{RC}^T = \mathbf{Y}^f$. This approach avoid to invert $\mathbf{R}$.
5. Compute the $k \times k$ matrix $\tilde{\mathbf{P}}^a = [(k-1)\mathbf{I}/\rho + \mathbf{CY}^f]^{-1}$. $\rho > 1$ is a multiplicative covariance inflation factor that is a remedy to "inflate" the forecast ensemble covariance that tends to be underweighted. The multiplication $\mathbf{CY}^f$ requires less that $2k^2l$ operations.
6. Compute the $k \times k$ matrix $\mathbf{W}^a = [(k-1)\tilde{\mathbf{P}}^a]^{\frac{1}{2}}$
7. Compute the k dimensional vector $\bar{\mathbf{w}} = \tilde{\mathbf{P}}^a \mathbf{C}(\mathbf{y} - \bar{\mathbf{y}}^f)$ and add it to each i column of $\mathbf{W}^a$ to form the analysis $k \times k$ weight matrix $\tilde{\mathbf{W}}^a$
8. Compute the analysis perturbation matrix $\mathbf{X}^a = \mathbf{X}^f \tilde{\mathbf{W}}^a$
9. The analysis ensemble $\mathbf{x}^{a_i}$, is formed by adding $\bar{\mathbf{x}}^f$ to the ith column of $\mathbf{X}^a$ with $i = 1, 2, \ldots, k$

The data assimilation is complete when the global analysis ensemble $\mathbf{x}^{a_i}$ is formed, which consists of the collection of local analysis ensemble at the center of each local region.

5.3 Mars Data Assimilation: The General Circulation Model

Because the data assimilation requires a reference model, this section is divided into two parts, one dealing with a Mars general circulation model and another that deals with Mars data assimilation .

Some planetary models derive from the Mintz-Arakawa' s model for Earth atmospheric circulation. There are several versions, from the first one up to the current UCLA's model (version 6.4) pass 40 years of research work. Akio Arakawa has been a leader in the field of Earth atmospheric general circulation model (AGCM) development from its beginning. AGCMs are essential tools for studies of global warming and projecting the consequences of anthropogenic climate change. His AGCMs contain several contributions on several areas

1. numerical schemes suitable for the long model integrations required by climate studies;
2. modeling of cloud processes including cumulus parameterization;
3. modeling of planetary boundary layer (PBL) processes.

The equations for large-scale atmospheric motion, the so called primitive equations, are defined by:

1. The hydrostatic equations
$$\frac{\partial \Phi}{\partial \sigma} = -b\mathbf{RT} \quad (5.55)$$
where the sigma coordinate ($\sigma \equiv \frac{p-p_T}{p_s-p_T}$) is the vertical coordinate, increasing downward, with p the pressure and p_s and p_T are respectively the pressure at the lower and upper boundaries of the atmospheric domain. Φ is the geopotential of the sigma coordinate surface, R is the gas constant, T is the temperature and $b \equiv (\sigma + \frac{p_T}{\pi})^{-1}$ with $\pi \equiv (p_s - p_T)$ where p_s is a a function of Φ, λ are the latitude and the longitude respectively, t is the time and p_T is a constant.
2. The horizontal momentum
$$\frac{\partial(\pi\mathbf{v})}{\partial t} = -div_h(\pi\mathbf{vv}) - \frac{\partial(\pi\dot{\sigma}\mathbf{v})}{\partial\sigma} - (2\Omega + \frac{u}{a\cos\Phi})(\mathbf{k}\times\pi\mathbf{v})$$
$$\sin\phi - [grad_h(\pi\Phi) - (\Phi - \sigma b\mathbf{RT})grad_h(\pi)] + \pi\mathbf{F} \quad (5.56)$$
where Ω is the planetary rotation rate, a is the planetary radius, $\mathbf{v}$ is the wind velocity with the eastward component u and westward v, ϕ is the latitude, λ is the longitude and t the time. $\mathbf{F}$ is the horizontal frictionless force per unit mass and $\mathbf{k}$ is the vertical unit vector, and div_h and $grad_h$ are respectively the divergence and gradient operators on the surface of constant σ.
3. The thermodynamic equation
$$\frac{\partial(\pi\mathbf{T})}{\partial t} = -div_h(\pi\mathbf{vT}) - \frac{\partial(\pi\dot{\sigma}\mathbf{T})}{\partial t} + C_p^{-1}b\mathbf{RT}\frac{dp}{dt} + C_p^{-1}\pi\mathbf{H} \quad (5.57)$$
where $\mathbf{H}$ is the rate of heating per unit mass, with the auxiliary relation which follows from the equation of continuity; C_p is the specific heat at constant pressure. Here $R = 0.188 \times 10^7 erggm^{-1}\mathrm{K}^{-1}$; $C_p = 0.87910^7 \times (1.0 + 0.634 \times 10^{-3}(T - 250))$ for T ranging from 120 to 300 K.
$$\pi\dot{\sigma} = -\int_0^{\sigma} div_h(\pi\mathbf{v})d\sigma - \sigma\frac{\partial\pi}{\partial t} \quad (5.58)$$
and the pressure tendency equation
$$\frac{\partial\pi}{\partial t} = -\int_0^1 div_h(\pi\mathbf{w})d\sigma - (\pi\dot{\sigma})_{\sigma=1} \quad (5.59)$$
and the substantial derivative of pressure
$$\frac{dp}{dt} = \sigma[\frac{\partial\pi}{\partial t} + \mathbf{w}\cdot grad_h(\pi)] + \pi\dot{\sigma}. \quad (5.60)$$

Mars general circulation modeling began in the 1960 s when Leovy and Mintz [18] modified the UCLA two level atmospheric model developed by Mintz [19] and Arakawa [20] for conditions appropriate for Mars, and used it to study the planet's wind systems, thermal structure, and energetics. More information on Mars geological structure and a description of the present atmosphere can be found in two books: *Mars* [21] and *An Introduction to its Interior, Surface and Atmosphere* [22].

The value of $\dot{\sigma} = 0$ at the lower boundary , where $\sigma = 1$ except when CO_2 condenses or sublimes. It is assumed $\frac{dp}{dt} = 0$ at $p = p_T$ so that $\dot{\sigma} = 0$ where $\sigma = 0$. The heating rate is the combined effect of solar heating, IR radiative transfer, convective exchange and latent heat due to condensation and sublimation of CO_2 which is the main gas on Mars.

Thus the rate of heating for unit mass is:

$$H_i = \frac{g}{(\Delta p)_i}(\Delta S_i + \Delta W_i + \Delta C_i) \tag{5.61}$$

where ΔS_i, ΔW_i, ΔC_i are the differences between the net flux at the top and bottom of the layer due to solar radiation, thermal radiation and sub-grid convection. Leovy and Mintz [18] report the equations to compute these quantities. The mass of the layer is represented by $\frac{(\Delta p)_i}{g}$ where $(\Delta p)_i$ is the increase in pressure from the top to the bottom of the layer. In the region where the pressure is lower than p_T they assume that it absorb heat.

Following the Mintz [19] and Arakawa [20] two layer Earth atmosphere model they subdivided the Mars atmosphere in the upper layer with $i = 1$ and lower layer with $i = 3$ where $(\Delta p)_1 = \frac{(p_s + p_T)}{2}$ and $(\Delta p)_3 = \frac{(p_s - p_T)}{2}$ respectively. The heating H_i is defined as:

1. Solar heating
   Along the CO_2 near IR absorption band, the differences can be formulated as:

$$\Delta S_1 = (\frac{r_m}{r})^2 (\sin\alpha)^{\frac{1}{2}} \times \{465 + [2397 + 531 \ln(\csc\alpha)](\sin\alpha)^{\frac{1}{2}}\} \tag{5.62}$$

   and

$$\Delta S_3 = (\frac{r_m}{r})^2 (\sin\alpha)^{\frac{1}{2}} \times \{378 + 657(\sin\alpha)^{\frac{1}{2}}\}, \tag{5.63}$$

   where $\frac{r_m}{r}$ is the ratio of the Mars mean distance to its actual distance and α is the solar elevation angle.
2. Infrared heating
   The flux differences in the IR 15 μ region, where is located the main infrared band of CO_2 are:

$$\Delta W_1 = -1.473 \times 10^6 Y(T_T) + [1.055 T_2 - 1367 - 16580 T_2^{-1}] \times [T_1 - T_3] \\ - 0.1282 \times 10^6 [Y(T_4) - Y(T_G)] \tag{5.64}$$

and

$$\begin{aligned}\Delta W_3 = &-0.455 \times 10^6 Y(T_T) + [1.50T_2 - 171 - 51530T_2^{-1}] \times [T_1 - T_3] \\ &- 1.80 \times 10^6 [Y(T_4) - Y(T_G)] + 1.302 \times 10^8 [Y(T_4)/T_4]^2 \\ &\times (T_4 - T_G) \exp(964.1/T_4), \end{aligned} \quad (5.65)$$

where

$$Y(T_k) = [\exp(964.1/T_k) - 1] \quad (5.66)$$

and T_4, T_2 are obtained from T_1 and T_3 by linear interpolation and extrapolation.

3. Convection
   Sub-grid convective heat transport at the ground is:

$$C_4 = \rho C_p C_H(X) u * (T_G - T_4), \quad (5.67)$$

where ρ is the mean surface air density, u^* is the surface friction velocity and $C_H(X)$ is a convective heat transfer coefficient which depends on $X = [\kappa g(T_G - T_4)/\bar{T}]^{\frac{1}{2}}/u^*$. κ is the molecular thermal diffusivity at the surface and $\bar{T}$ is the global mean surface air temperature. For stable conditions one assumes $C_H(X) = C_M$ where C_M^2 is the momentum drag coefficient for stable conditions. When the atmosphere is unstable one uses an empirical formula:

$$C_H(X) = k_0 / \int_0^\infty [(X^3 C_H/\zeta_T) + \zeta/\psi(\zeta)]^{-1} d\zeta, \quad (5.68)$$

where $\psi(\zeta)$ is an empirical function, ζ_T is an empirical parameter and k_0 is the von Karman constant. The value of C_H can assume the following values

$$\begin{cases} C_H(X) = 0.142X^{0.14} & \text{if } X \leq 0.153 \\ C_H(X) = 0.204X^{\frac{1}{3}} & \text{if } X > 0.153. \end{cases}$$

When the lapse rate is unstable ($\gamma > \gamma_a$) the upward convective heat flux at level 2 is the rate

$$C_2 = 4 \times 10^2 T_2(\gamma - \gamma_a), \quad (5.69)$$

where γ is the actual lapse rate between T_1 and T_3 and $\gamma_a = \frac{g}{C_p}$ is the adiabatic lapse rate. $\gamma_a = 4.23 \times (1.0 + 0.634 \times 10^{-3}(T_2 - 250))$ K Km^{-1}. When the lapse rate is stable but exceeds the threshold $\gamma_m \equiv 2.5$ Km^{-1} and when the surface heat flux C_4 is upward, a convective exchange arises. In such a case

$$C_2 = C_4 \frac{\gamma - \gamma_m}{\gamma_a - \gamma_m}. \quad (5.70)$$

4. Surface heat balance
   From the heat balance equation

$$(1 - A)S_4 - W_4 - C_4 - D - L = 0, \tag{5.71}$$

   one can obtain the ground temperature T_G. In it A is the surface albedo, S_4 is the downward solar radiation, W_4 is the net upward IR radiation at the surface, C_4 is the net upward convective heat flux at the surface, D is the downward conductive heat flux into the soil and L is the latent heat release due to CO_2 condensation on the surface.
5. Friction and lateral diffusion
   Since some of the thermal energy introduced into the system is converted into kinetic energy of the wind field that is dissipated by the frictional force F linked with the vertical eddy stress, for the two layer model by Leovy and Mintz [18] one has:

$$F_1 = \frac{2g}{p_s - p_T}\tau_2 \tag{5.72}$$

   and

$$F_2 = \frac{2g}{p_s - p_T}(\tau_2 - \tau_s) \tag{5.73}$$

   where $\frac{p_s - p_T}{2g}$ is the mass per unit area of each layer and τ_2 is the stress on the lower layer due to turbulent exchange

$$\tau_2 = \frac{p_s - p_T}{2g}C^*(v_1 - v_2) \tag{5.74}$$

   where C^* is a parameter depending on the lapse rate. The surface stress τ_s is

$$\tau_s = \rho_s u *^2 (\frac{v_s}{|v_s|}) \tag{5.75}$$

   where $u* = C_M |v_s|$, with v_s the surface wind and the momentum drag coefficient is C_M.

   Since in Earth simulation model there also a lateral diffusion Leovy and Mintz also introduced a Mars lateral diffusion coefficients depending on the local grid distance.

Several sophisticated Mars global circulation models (GCMs) have been developed since. As an example among these, Richardson et al. [23] derived Planet-WRF from WRF, which had been designed for modelling terrestrial meso-scale and micro-scale. WRF was modified to work as a planetary model (see http://planetwrf.com/model) and the main changes are:

1. Non-Conformal Grid Modification.
   Instead of using the conformal map of WRF where the scale factors, that relates map projection distance to physical distance are independent of direction (as happens when one uses polar stereographic, Lambert conformal, Mercator, and simplified/idealized square grid boxes), the PlanetWRF was generalized so that non-conformal map projections could be used. The equations were written with separate x- and y-directional map scale factor components. These choices allow any generalized map projection to be used, including transverse projection and stretched/zoomed grids. Thus the map projection of planetWFR was built to run on a simple cylindrical projection, evenly spaced in latitude and longitude.
2. Polar boundaries and filtering.
   In the simple cylindrical map projection the grid points can be defined to cover the entire globe in latitude and longitude. In the east-west direction, where there is a flow across the border and onto the opposite side, the appropriate boundary conditions are periodic. In the north-south direction a polar boundary condition is needed. PlanetWRF has adopted the convention for C-grid models with the standard sigma coordinate. The Arakawa grid system represents and computes orthogonal physical quantities as the velocity and mass-related quantities, on rectangular grids used for Earth system models in meteorology and oceanography. The five Arakawa grids (A-E) were first introduced in Arakawa and Lamb [24]. The "staggered" Arakawa C-grid , instead of evaluating both east-west (u) and north-south (v) velocity components at the grid center, separates the vector quantities evaluating the u components at the centers of the left and right grid faces, and the v components at the centers of the upper and lower grid faces. PlanetWRF takes the polar point to be a C-grid v-stagger point (see Fig. 5.1), with the value of

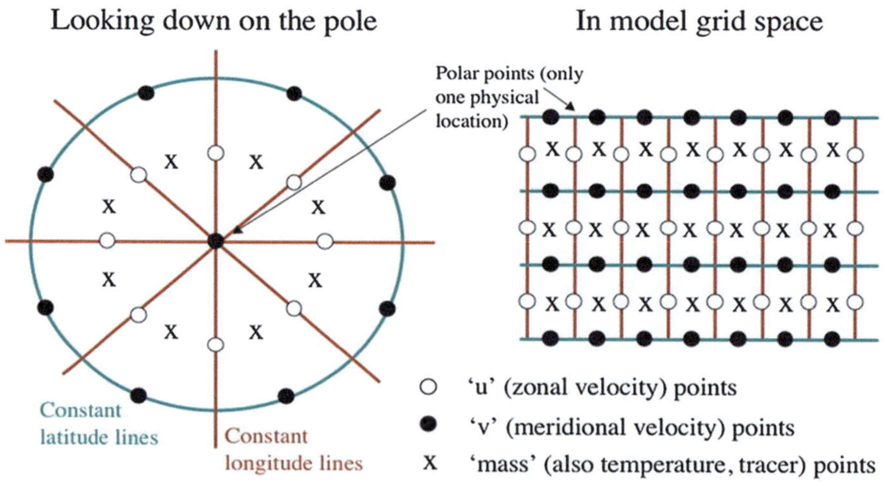

Fig. 5.1 PlanetWRF global grid. Looking down on the pole (*left*). C-grid cylindrical projection (*right*). The grid points are equally spaced in longitude and latitude with horizontally staggered u, v and mass points. At the pole all v points correspond to one physical location. source Richardson et al. [23]

the meridional velocity constantly equal to zero. Flux and gradient calculations across the pole are not allowed. This does not preclude advection of material across the pole. The advection over the poles is instead accomplished by zonal transport within the most poleward zone.

Since the physical distance for zonal advection of information decreases rapidly to zero at the pole, the model time step needed to avoid instabilities in the horizontal direction. In numerical analysis, when certain partial differential equations (usually hyperbolic PDEs) are solved, the time step must be less than a certain threshold, otherwise the simulation will produce incorrect results. The Courant-Friedrichs-Lewy [25] or CFL criteria is a necessary condition for convergence by the method of finite differences. To be able to use longer time steps (those suitable for more tropical latitudes) PlanetWRF has implemented polar Fourier filtering of the higher-frequency components of state variables. All grid points poleward of 60 degrees are filtered, with the cutoff frequency being a function of the cosine of latitude. To yield the greatest stability, the column mass, horizontal winds, temperature and tracers (moisture, aerosols, chemicals, etc.) are also filtered

3. Generalized Planetary Parameters and Timing Conventions.
All planet-specific parameters such as orbital parameters, the relationship between MKS (SI) seconds and model seconds defined as 1/86400 of the rotational period of a planet, reference pressure, gravity and gas constant are set in one centralized module where there are also the set of consistent planetary parameters.

   Since the model assumes that one day is made up of 24 h, each of which is composed of 60 minutes, in turn made up of 60 s, then there is always an integer number of time steps per day. However, the dynamics and physics routines are still integrated in MKS (SI) units, with the conversion from model-to-SI time made before calculating tendencies and physical quantities. Other items are:

   a. Since WRF version uses the standard day-month-year calendar format, this convention is used to drive the solar radiation routines and to label model output.
   b. Since WRF uses routines from the standardized Earth System Modeling Framework (ESMF) planetWRF converts these routines to drive the model with user-specified orbital elements using the planetocentric solar longitude (L_s) date system (with $L_s = 0$ corresponding to northern hemisphere spring equinox and $L_s = 90$ to northern hemisphere summer solstice, etc.).

4. Parameterizations of sub-grid scale physical processes for various planet.
Depending on the planet, physical routines have been added to WRF to treat the radiative transfer in atmosphere with high aerosol optical thickness and CO_2 atmospheric gas, also taking into account the condensation/sublimation of CO_2 from polar caps (Mars), For Titan, an updated version of the radiative transfer scheme described by McKay et al. [26] has been developed. Simple surface drag and radiative relaxation schemes, similar to the Held and Suarez [27] forcing has been used to validate the global dynamical core and to treat the Venusian atmosphere. The concept of dynamical core was introduced to extend the ideas of plug-compatible parameterizations to the coding of the dynamics. PlanetWRF

has the same WRF existing horizontal and vertical diffusion parameterization schemes, as far as the physics of diffusion remains the same with only the diffusivities varying.

5.3.1 Mars Data Assimilation: Methods and Solutions

To perform an assimilation Lee et al. [28] adopted version 3.0.1.1 of the MarsWRF climate model (Richardson et al., [23]) with a radiative forcing in the GCM using a two stream, single scattering, radiative flux solver based on the Hadley Centre Unified Model algorithm (Edwards and Slingo, [29]), modified by Mischna et al. [30]. This parameterization calculates fluxes in the visible and infrared spectra using a correlated k method to couple the optical properties of the carbon dioxide atmosphere with the Mie scattering [31] parameters in order to describe a radiative atmosphere where there are also dust and water and aerosols.

To create the ensemble of model states, the steady state atmosphere is perturbed using additive Gaussian white noise perturbations on the temperature, surface pressure, horizontal wind, emissivity, albedo, and column dust opacity. Each ensemble member is then integrated for a certain number of *sols* to reach a new steady state. The term *sol* is used by planetary astronomers to refer to the duration of a solar day on Mars. A mean Martian solar day, or *sol*, is 24 h, 39 min, and 35.244 s. The magnitudes of the perturbations are small (e.g. standard deviations of 2 K in temperature, 5 m/s in horizontal wind, 0.02 in albedo and emissivity) and are constrained to keep albedo and emissivity between zero and one.

At each assimilation step an observation forward model within the data assimilation system has been used to simulate the observations using the state vectors of the ensemble members. The simulated observations are the calibrated radiances observed by the Thermal Emission Spectrometer (TES) [32] aboard the Mars Global Surveyor (MGS). TES is a nadir sounding grating spectrometer, observing the spectrum from 200 to 1700 cm^{-1} with a resolution of 5 cm^{-1} or 10. cm^{-1} depending on the scan mode. These data can be obtained from the NASA Planetary Data System archive (PDS, http://pds.jpl.nasa.gov).

The integration step is performed by the MarsWRF GCM under the control of the DART software, which maintains the ensemble of model states and provides the initial conditions necessary to continue the simulation. The assimilation step is performed by the DART software.

DART is a facility for ensemble Data Assimilation developed and maintained by the Data Assimilation Research Section (DAReS) at the National Center for Atmospheric Research (NCAR) http://www.image.ucar.edu/DAReS/DART/. DART is a software that makes to explore a variety of data assimilation methods and observations with different numerical models and is designed to facilitate the combination of assimilation algorithms, models, and real (as well as synthetic) observations to allow increased understanding of all three.

DART employs a modular programming approach to apply an Ensemble Kalman Filter which nudges the underlying models toward a state that is more consistent with information from a set of observations. Models may be swapped in and out, as can different algorithms in the Ensemble Kalman Filter. The method requires running multiple instances of a model to generate an ensemble of states. A forward operator appropriate for the type of observation being assimilated is applied to each of the states to generate the model's estimate of the observation.

A sequential data assimilation scheme has been implemented and validated by Montabone et al. [33] and Lewis et al. [34]. They derived the analysis correction scheme for Mars, from the one used operationally for Earth weather forecasting at the UK Meteorological Office. The scheme has been interfaced with the state-of-the-art Mars General Circulation Model (MGCM) under development at the University of Oxford in the UK and at Laboratoire de Meteorologie Dynamique du CNRS in Paris, France. A reanalysis of almost complete Martian years has been carried out using retrieved thermal profiles and column dust optical depth by data from TES mapping phase and has been made publicly available.

Greybush et al. [35] have developed an Ensemble Mars Atmosphere Reanalysis System (EMARS) based on the LETKF scheme (Hunt et al., [17]) and the Geophysical Fluid Dynamics Laboratory (GFDL, Princeton, NY, USA) Mars GCM. Each ensemble member represents a potential atmospheric state, spanning a range of possibilities, with the ensemble mean being the most probable and the ensemble spread reflecting the uncertainty. The LETKF also has the ability to estimate and improve model parameters. They have assimilated retrievals of thermal profiles and column dust optical depth from TES during its entire mapping phase as well.

Most recently, Navarro et al. [36] have applied the same LETKF scheme used by Greybush et al. [35] to a different MGCM, i.e. the one maintained by the Laboratoire de Meteorologie Dynamique in Paris. They have also used different observations, i.e. the thermal profiles retrieved from the radiances measured by the Mars Climate Sounder aboard the Mars Reconnaissance Orbiter—the US spacecraft which followed MGS in the atmospheric mapping on Mars.

5.4 Earthquake Forecast

Earthquake statistics clearly violate Gaussian approximations in terms of their temporal, spatial and magnitude occurrences, so much so that approximate algorithms based on local Gaussian approximations (for instance the extended Kalman filter) are highly unlikely to produce good results. Furthermore, the continuous state space of seismicity rules out methods in which that space is assumed to be discrete such as grid-based methods.

Then the approach is with numerical integration techniques and Monte Carlo methods. The former are numerically accurate but computationally expensive in problems with high dimensionality, while the Sequential Monte Carlo (SMC) methods bridge the gap between these cost-intensive methods and the methods based on Gaussian approximations. They are a set of simulation based methods that provide a flexible alternative to computing posterior distributions. They are applicable in very general settings and often relatively easy to implement. SMC methods are also known under the names of particle filters, bootstrap filters, condensation, Monte Carlo filters, interacting particle approximations and survival of the fittest. Good introductions can be found, for instance in Arulampalam et al. [37],

As we have seen in previous chapters the Bayesian filtering covers different scenarios: from Kalman filter to Sequential Monte Carlo sampling, the so called particle filters. Among the Monte Carlo methods one of the simplest is the Importance Sampling (IS) that introduces the idea of generating samples from a known, easy-to-sample probability density function *pdf*, $q(x)$, called the importance density or proposal density, and then "correcting" the weights of each sample so that the weighted samples approximate the desired density. However IS, in its simplest form, is not adequate for sequential estimation. Whenever new data become available, one needs to recompute the importance weights over the entire state sequence.

Sequential Importance Sampling (SIS) modifies IS so that it becomes possible to compute an estimate of the posterior without modifying the past simulated trajectories. The problem encountered by the SIS method is that, as time increases, the distribution of the importance weights becomes more and more skewed. For instance, if the support of the importance density is broader than the posterior density, then some particles will have their weights set to zero in the update stage. But even if the supports coincide exactly, many particles will over time decrease in weight so that after a few time steps, only a few lucky survivors have significant weights, while a large computational effort is spent on propagating unimportant particles.

It has been shown that the variance of the weights can only increase over time, thus it is impossible to overcome the degeneracy problem (Kong et al., [38]). Two solutions exist to minimize this problem:

1. a good choice of the importance density,
2. resampling.

Sequential Importance Resampling (SIR) introduces, however, other problems. Since particles are sampled from discrete approximations to density functions, the particles with high weights are statistically selected many times. This leads to a loss of diversity among the particles as the resultant sample will contain many repeated points. This is known as "sample impoverishment" (see Arulampalam et al. [37]) and is severe when the model forecast is very narrow. There are various methods to deal with this problem, including sophisticated methods that recalculate past states and weights via a recursion and Markov Chain Monte Carlo (MCMC) methods. Because of the additional problems introduced by resampling, it makes sense to resample only when the variance of the weights has decreased appreciably.

In the next section we introduce the Earthquake model and the Sequential Monte Carlo and related methods applicable to the Earthquake events.

5.4.1 Renewal Process as Forecast Model

One of the model used in seismology is the renewal processes. In general the physics model that are used for seismic hazard map are not well defined and are not unique, and the same renewal processes models, which are motivated from the elastic rebound proposed by Reid [39], are under discussion. According to the theory, a large earthquake release the elastic strain that has been built up since the last large earthquake.

Then renewal point process are characterized by interval between successive events that are identically and independently distributed according to a probability density function that defines the process (Daley and Vere-Jones [40]).

The time of the next event only depends on the time of the last event:

$$p(t_k|t_{k-1}) = p(t_k - t_{k-1}) = p(\tau) \tag{5.76}$$

where τ represent the interval between the two events. In this frame the time of the event t_k corresponds, in data assimilation, to the model state. The renewal point processes provides the prior information for the analysis.

Werner et al., [41] offers a pedagogical example of earthquake forecasting. Their model is based on lognormal renewal process [42, 43].

The intervals of τ of the lognormal process are distributed according to:

$$f(\tau, \mu, \sigma) = \frac{1}{\tau\sqrt{2\pi\sigma}} \exp(-\frac{\log\tau - \mu^2}{2\sigma^2}) \tag{5.77}$$

where the parameter μ and σ may need to be estimated.

In terms of lognormal distribution we have

$$p(t_k - t_{k-1}, \mu, \sigma) = \frac{1}{(t_k - t_{k-1})\sqrt{2\pi\sigma}} \exp(-\frac{\log(t_k - t_{k-1}) - \mu^2}{2\sigma^2}) \tag{5.78}$$

with $\sigma = 0.7$ and $\mu = -0.245$ given by Biasi et al. [44]. Since the observed occurrence time t_k^o is given by the "true" occurrence time t_k^t affected by the noise ϵ_k we can write:

$$t_k^o = t_k^t + \epsilon_k \tag{5.79}$$

Werner et al. [41] run the numerical experiments using two different distribution.

1. Gaussian uniform distribution
   It is:

$$p_{uniform}(\epsilon) = \frac{1}{\Delta} H(\epsilon + \frac{\Delta}{2}) H(\frac{\Delta}{2} - \epsilon) \tag{5.80}$$

$$p_{uniform}(\epsilon) = \begin{cases} \frac{1}{\Delta} & -\frac{\Delta}{2} \le \epsilon \le +\frac{\Delta}{2} \\ 0 & \text{otherwise} \end{cases}$$

where $H(\cdot)$ is the Heaviside step function. The conditional likelihood is obtained substituting $\epsilon = t^o - t_k^t$

$$p_{uniform}(\epsilon_k) = p(t_k|t_{k-1}) = p(t_k^o - t_k^t) \tag{5.81}$$

$$p_{uniform}(\epsilon) = \begin{cases} \frac{t_k}{\Delta} & t_k - \frac{\Delta}{2} \le t_k^o \le t_k + \frac{\Delta}{2} \\ 0 & \text{otherwise} \end{cases}$$

where the parameter $\Delta = 0.5$

2. The Gaussian mixture model is

$$p_{mix}(\epsilon) = p_1\mathcal{N}(\eta_1, \rho_1) + p_2\mathcal{N}(\eta_2, \rho_2) \tag{5.82}$$

where $p_1 = 0.4$ and $p_2 = 0.6$ and the normal distribution $\mathcal{N}$ is characterized by $\eta_1 = \pm 0.2$, $\sqrt{\eta_2} = 0.02$, $\sqrt{\eta_2} = 0.01$

The simulation is obtained generating the true estimate by n random samples from lognormal distribution (5.78).

The particle filtering was initialized with $N = 10000$ particles at t_0 and the forecast t_1 was obtained propagating each particle though the model (5.78). Once they had obtained the observation estimate and the model forecast the SIR particle filter (described in the next paragraph) was used to obtain the analysis at t_1. Then posterior approximation is used to forecast t_2 by the marginal distribution of the recursive bayesian filter.

It is interesting to note that Werner et al. [41] make the comparison between SIR and EnSRF obtaining quite similar results in the case of uniform Gaussian noise but fail with problems in which the measurements errors are non Gaussian.

5.4.2 Sequential Importance Sampling and Beyond

Sequential importance sampling algorithm introduced by Marshall [45] is a Monte Carlo (MC) method that forms the basis for most sequential MC sampling approximation known as bootstrap filtering, jackknife techniques, condensation algorithm, particle filtering etc. It is a technique for implementing a recursive Bayesian filter by MC simulations. The key idea (Arulampalam et al. [37]) is to represent the required a posteriori density function by a set of random samples with associated weights.

Let us consider a state sequence

$$\mathbf{x}_k = \mathbf{f}_k(\mathbf{x}_{k-1}, \mathbf{w}_{k-1}) \tag{5.83}$$

and the measurements

$$\mathbf{y}_k = \mathbf{h}_k(\mathbf{x}_k, \mathbf{v}_k) \tag{5.84}$$

where $\mathbf{f}_k$ and $\mathbf{h}_k$ are possibly non linear functions and $\mathbf{w}_{k-1}$ and $\mathbf{v}_k$ are independent and identical distributed (i.i.d.) process and measurement noise sequences respectively. The problem is to seek filtered estimates of $\mathbf{y}_k$ based on the set of all available measurements $\{\mathbf{y}_i, \quad i = 1, \ldots, k\}$ as we have already seen in the bayesian chapter.

From a Bayesian point of view the tracking problem is to construct the probability density function pdf, $p(\mathbf{x}_{0:k}|y_{1:k})$ that can recursively be obtained in two stages: forecast and update. The initial pdf is: $p(\mathbf{x}_0|z_0) \equiv p(\mathbf{x}_0)$ as we have seen previously.

When the true density is non-Gaussian, the approximated grid based filters and particle filters will improve the performance respect to other approaches that require Gaussian approximation.

In the Monte Carlo sampling an empirical posterior density at k can be expressed as

$$p(\mathbf{x}_{0:k}|y_{1:k}) = \sum_{i=1}^{N_p} \delta(\mathbf{x}^i_{0:k}) \tag{5.85}$$

where δ is the Delta Dirac mass located at $\mathbf{x}^i_{0:k}$. Considering a function $f(\mathbf{x}_{0:k})$ integrable in a measurable space, the Lebesque-Stieltjes integral is the estimate of such function

$$I_{N_p}(f) = \int f(\mathbf{x}_{0:k})p(\mathbf{x}_{0:k}|y_{1:k})d\mathbf{x}_{0:k} = \frac{1}{N_p}\sum_{i=1}^{N_p} f(\mathbf{x}^i_{0:k}), \tag{5.86}$$

that is unbiased.

In the importance sampling the objective is to sample the distribution in the region of "importance" in order to achieve computational efficiency. This is important especially for the high-dimensional space where the data are usually sparse, and the region of interest where the target lies in is relatively small respect to the whole data space.

Thus let denote a random measure that is characterized by the posterior pdf, $p(\mathbf{x}_{0:k}|z_{1:k})$, where $\{\mathbf{x}^i_{0:k}, \quad i = 1, 2, \ldots, N_p\}$ is a set of support points with associated weights $\{\mathbf{w}^i_k = \frac{p(x^i)}{q(\mathbf{x}^i)}, \quad i = 1, 2, \ldots, N_p\}$ and $\{\mathbf{x}^i_{0:k}, \quad i = 1, 2, \ldots, N_p\}$ is the set of all states up to the time k.

One chooses a proposal distribution $q(\mathbf{x}_{0:k}|\mathbf{y}_{1:k})$ in place of a true probability distribution $p(\mathbf{x}_{0:k}|\mathbf{y}_{1:k})$ so that

$$p(\mathbf{x}_{0:k}|\mathbf{y}_{1:k}) = \frac{p(\mathbf{x}_{0:k}|\mathbf{y}_{1:k})}{q(\mathbf{x}_{0:k}|\mathbf{y}_{1:k})}q(\mathbf{x}_{0:k}|\mathbf{y}_{1:k}), \tag{5.87}$$

to obtain the identity

$$I(f) = \frac{\int f(\mathbf{x}_{0:k})\mathbf{w}(\mathbf{x}_{0:k})q(\mathbf{x}_{0:k}|\mathbf{y}_{1:k})d\mathbf{x}_{0:k}}{\mathbf{w}(\mathbf{x}_{0:k})q(\mathbf{x}_{0:k}|\mathbf{y}_{1:k})d\mathbf{x}_{0:k}}, \tag{5.88}$$

where

$$\mathbf{w}(\mathbf{x}_{0:k}) = \frac{p(\mathbf{x}_{0:k}|\mathbf{y}_{1:k})}{q(\mathbf{x}_{0:k}|\mathbf{y}_{1:k})}. \tag{5.89}$$

When one generates N identically and independently distributed samples $\mathbf{x}_{0:k}^i$ from the importance sampling $q(\mathbf{x}_{0:k}|\mathbf{y}_{1:k})$, a Monte Carlo estimate of $I(f)$ is

$$\hat{I}(f) = \frac{\frac{1}{N_p}\sum_{i=1}^{N_p} f(\mathbf{x}_{0:k}^i)\mathbf{w}(\mathbf{x}_{0:k}^i)}{\frac{1}{N_p}\sum_{j=1}^{N_p}\mathbf{w}(\mathbf{x}_{0:k}^j)} = \sum_{i=1}^{N_p} f(\mathbf{x}_{0:k}^i)\tilde{\mathbf{w}}_k^i, \tag{5.90}$$

where the normalized importance weights $\tilde{w}_k^i$ is

$$\tilde{\mathbf{w}}_k^i = \frac{\mathbf{w}(\mathbf{x}_{0:k}^i)}{\sum_{j=1}^{N_p}\mathbf{w}_{0:k}^j}. \tag{5.91}$$

Then the posterior density is

$$\hat{p}(\mathbf{x}_{0:k}|\mathbf{y}_{1:k}) = \sum_{i=1}^{N_p}\tilde{\mathbf{w}}_k^i\delta(\mathbf{x}_{0:k}^i). \tag{5.92}$$

The importance sampling may be useful to reduce the variance of the estimator and when encountering the difficulty to sample from the true probability.

In general, however, it is difficult to find a good proposal distribution function especially in a high dimensional space. If the importance density $q(\mathbf{x}_{0:k}|\mathbf{y}_{1:k})$ at time k admits a marginal distribution at time $k-1$, the importance function $q(\mathbf{x}_{0:k}|\mathbf{y}_{1:k-1})$ can construct the proposal sequentially, which is the basis of Sequential Importance Sampling (SIS)

$$q(\mathbf{x}_{0:k}|\mathbf{y}_{1:k}) = q(\mathbf{x}_{0:k-1}|\mathbf{y}_{1:k-1})q(\mathbf{x}_k|\mathbf{x}_{0:k-1},\mathbf{y}_{1:k}) \tag{5.93}$$

After iteration, the proposal is obtained in a factorized form

$$q(\mathbf{x}_{0:k}|\mathbf{y}_{1:k}) = q(\mathbf{x}_0)\prod_{l=1}^{k} q(\mathbf{x}_{0:l}|\mathbf{x}_{0:l-1},\mathbf{y}_{1:l}) \tag{5.94}$$

and the importance sampling can be performed recursively.

Assuming the state evolves according to a Markov process and the observation are conditionally independent

$$p(\mathbf{x}_{0:k}) = p(\mathbf{x}_0)\prod_{l=1}^{k} p(\mathbf{x}_l|\mathbf{x}_{l-1}) \tag{5.95}$$

and

$$p(\mathbf{y}_{1:k}|\mathbf{x}_{0:k}) = \prod_{l=1}^{k} p(\mathbf{y}_l|\mathbf{x}_l) \tag{5.96}$$

According to the law of probability, the importance weights $\tilde{\mathbf{w}}_k^i$ given by the relation (5.91), and using Eqs. (5.94), (5.95), (5.96), can be written as

$$\tilde{\mathbf{w}}_k^i \propto \tilde{\mathbf{w}}_{k-1}^i \frac{p(\mathbf{y}_k|\mathbf{x}_k^i)p(\mathbf{x}_k^i|\mathbf{x}_t^i|\mathbf{x}_{k-1}^i)}{q(\mathbf{x}_k^i|\mathbf{x}_{0:k-1}^i, \mathbf{y}_{1:k})} \tag{5.97}$$

This equation provides the importance weights for the sequential updating. The advantage of SIS is that it does not rely on the underlying Markov chain. The disadvantage is that the importance weights may have large variances and there is a problem of degeneracy. In order to minimize the problem there are two possible solutions. The first is to choose a good importance density, while the second is the resampling.

The sampling-importance resampling (SIR) is motivated from the bootstrap and jackknife techniques. The bootstrapping is to evaluate the properties of an estimator through the empirical cumulative distribution function (*cdf*) of the samples instead of the true *cdf*.

In statistics, the jackknife is a resampling techniques and bootstrapping is the evaluation of the properties of an estimator through the empirical cumulative distribution function (Efron and Tibshirani [46]).

In order to overcome the degeneracy a suitable method is to compute the effective sample size N_{eff} (Liu and Chen [47]) defined as

$$N_{eff} = \frac{N_p}{1 + Var(\mathbf{w}_k^{*i})} \tag{5.98}$$

where

$$\mathbf{w}_k^{*i} = \frac{p(\mathbf{x}_k^i|\mathbf{y}_{1:k})}{q(\mathbf{x}_k^i|\mathbf{x}_{k-1}^i, \mathbf{y}_k)} \tag{5.99}$$

When this may be not available and an estimate $\hat{N}_{eff}$ can be obtained as an inverse of the so called Participation Ratio (Mézard et al. [48])

$$\hat{N}_{eff} = \frac{1}{\sum_{i=1}^{N_p} (\mathbf{w}_k^i)^2} \tag{5.100}$$

Thus resampling is applied when $\hat{N}_{eff}$ falls below a certain threshold.

5.4.3 The Receipt of SIR

The SIR receipt is

1. Draw N_p random samples $\{\mathbf{x}^i, \quad 1 = 1, \ldots, N_p\}$ from proposal distribution $q(\mathbf{x})$;
2. Calculate importance weights $\mathbf{w}^i$ for each sample $\mathbf{x}^i$;
3. Normalize the importance weight to obtain $\tilde{\mathbf{w}}^i$;
4. Resample with replacement N times from the discrete set $\{\mathbf{x}^i, \quad 1 = 1, \ldots, N_p\}$ where the probability of resampling from $\mathbf{x}^i$ is proportional to $\tilde{\mathbf{w}}^i$;
5. Calculate $\hat{N}_{Jeff}$, if $\hat{N}_{Jeff} < N_{threshold}$ resample.

References

1. Lorenz, E.N.: Designing chaotic models. J. Atmos. Sci. **62**, 1574–1587 (2005)
2. Lorenz, E.N.: Atmospheric Predictability. Advances in Numerical Weather Prediction, 1965–66 Seminar Series. Travelers Research Center, Inc., pp. 34-39 (1966)
3. Lorenz, E.N., Emanuel, K.A.: Optimal sites for supplementary weather observations: simulation with a small model. J. Atmos. Sci. **55**, 399–414 (1998)
4. Lorenz, E.N.: Deterministic non-periodic flow. J. Atmos. Sci. **20**, 130–141 (1963)
5. Salzman, B.: Finite amplitude free convection as an initial value problem I. J. Atmos. Sci. **19**, 329–341 (1962)
6. Lyapunov, A.M.:The general problem of the stability of motion, Translated by Fuller, A.T. Taylor and Francis, London. ISBN: 978-0-7484-0062-1 Reviewed in detail by Smith, M.C.: Automatica 1995 **3**(2), 353–356 (1992)
7. McLaughlin, J.B., Martin, P.C.: Transition to turbulence of a statically stressed fluid system. Phys. Rev. A **12**, 186 (1975)
8. Glatzmaier, G.A., Roberts, P.H.: A three-dimensional self-consistent computer simulation of a geomagnetic field reversal. Nature **377**, 203–209 (1995)
9. Press, W.H., Flannery, B.P., Teukolsky, S.A., Vetterling, W.T.: Numerical Recipes. Cambridge University Press, Cambridge (1986)
10. Kuhl, D., Kostelich, E.: Introduction to LETKF Data Assimilation(2008).https://www.atmos.umd.edu/~dkuhl/AOSC614/LETKF_lab.pdf
11. Bannister, R.: A square root ensemble Kalman filter demonstration with the Lorenz model (2012). https://www.met.reading.ac.uk/textdarc/training/lorenze_nsrkf/
12. Migliorini, S.: Ensemble data assimilation with the lorenz equations (2010). https://www.met.reading.ac.uk/~hraa/projects/05/index.html
13. Chaplain, M.: Modelling Aspects of Cancer Growth: Insight from Mathematical and Numerical Analysis and Computational Simulation. Multiscale Problems in the Life Sciences Lecture Notes in Mathematics **1940**, 147–200 (2008)
14. Swanson, K.R., Alvord Jr, E.C., Murray, J.D.: A quantitative model of differential motility of gliomas in white and grey matter. Cell Prolif. **33**, 317–329 (2000)
15. Giatili, S.G., Stamatakos, G.S.: A detailed numerical treatment of the boundary conditions imposed by the skull on a diffusion-reaction model of glioma tumor growth. Clinical validation aspects. Appl. Math. Comput. **218** (2012)
16. Kostelich E.J., Kuang, Y., McDaniel, J.M., Moore, N.Z., Martirosyan, N.L., Preul, M.C.: Accurate state estimation from uncertain data and models: an application of data assimilation to mathematical models of human brain tumors, Biol. Direct 1, 6-64 (2011). http://www.biology-direct.com/content/6/1/64
17. Hunt, B.R., Kostelich, E.J., Szunyogh, I.: Efficient data assimilation for spatiotemporal chaos: a local ensemble transform Kalman filter. Physica D **230**, 112–126 (2007)

18. Leovy, C., Mintz, Y.: The numerical simulation of atmospheric circulation and climate of Mars. J. Atmos. Sci. **26**(6), 1167–1190 (1969)
19. Mintz, Y.: Very long-term global integration of the primative equations of atmospheric motion. (An experiment in climate simulation) WMO tech notes No. 66 141–167, also Meteorol. Monogr. **8**(30), 20–36 (1965)
20. Arakawa A.: Numerical simulation of large-scale atmospheric motions. Numerical solution of field problems continuum physics. In: Birkhoff, G., verge. S. (eds,) American Math Society, vol 2, pp 24-40. Providence R.I. (1970)
21. Matthews, M.S., Kieffer, H.H., Jakosky, B.M., Snyder, C.: Mars. The University of Arizona Press
22. Barlow, N.: Mars: An Introduction to its Interior, Surface and Atmosphere. Cambridge Planetary Science. Cambridge University Press, Cambridge (2014)
23. Richardson, M.I., Toigo, A.D., Newman, C.E.: PlanetWRF: A general purpose, local to global numerical model for planetary atmospheric and climate dynamics. J. Geophys Res. **112** (2007)
24. Arakawa, A., Lamb, V.R.: Computational Design of the Basic Dynamical Processes of the UCLA General Circulation Model. Methods of Computational Physics 17, pp. 173–265. Academic Press, New York (1977)
25. Courant R., Friedrichs, K., Lewyt, H.: On the Partial Difference Equations of Mathematical Physics. IBM J., (1967)
26. McKay, C.P., Pollack, J.B., Courtin, R.: The thermal structure of Titan's atmosphere. Icarus **80**(1), 23–53 (1989)
27. Held, I.M., Suarez, M.J.: A proposal for the intercomparison of the dynamical cores of atmospheric general-circulation models. Bull. Am. Meteorol. Soc. **72**(10), 1825–1830 (1994)
28. Lee, C., Lawson, W.G., Richardson, M.I., Anderson, J.L., Collins, N., Hoar, T., Mischna, M.: Demonstration of ensemble data assimilation for Mars using DART, MarsWRF, and radiance observations from MGS TES. J. Geophys. Res. **116**, 1–17 (2011)
29. Edwards, J.M., Slingo, A.: Studies with a flexible new radiation code: 1. Choosing a configuration for a large scale model. Q. J. R. Meteorol. Soc. **122**(531), 689–719 (1996)
30. Mischna, M.A., Toigo, A.D., Newman, C.E., Richardson, M.I.: Development of a new global, scalable and generic general circulation model for studies of the Martian atmosphere. Paper presented at the Second Workshop on Mars Atmosphere Modelling and Observations, Granada, Spain, CNES, 27 February–3 March (2006)
31. Levoni, C., Cervino, M., Guzzi, R., Torricella, F.: Atmospheric aerosol optical properties : a database of radiation characteristic for different components and classes. Appl. Opt. **36** (1997)
32. Christensen, P.R., et al.: Mars Global Surveyor Thermal Emission Spectrometer experiment: Investigation description and surface science results, J. Geophys. Res., **106**(E10) (2001)
33. Montabone, L., Lewis, S.R., Read, P.L., Hinson, D.: Validation of Martian meteorological data assimilation for MGS/TES using radio occultation measurements. Icarus **185**, 113–132 (2006)
34. Lewis, S.R., Read, P.L., Conrath, B.J., Pearl, J.C., Smith, M.D.: Assimilation of Thermal Emission Spectrometer atmospheric data during the Mars Global Surveyor aerobraking period. Icarus **1932**(2), 327–347 (2007)
35. Greybush, S.J., Wilson, R.J., Hoffman, R.N., Hoffman, M.J., Miyoshi, T., Ide, K., McConnochie, T., Kalnay, E.: Ensemble kalman filter data assimilation of thermal emission spectrometer temperature retrievals into a Mars GCM. J. Geophys. Res. **117** (2012)
36. Navarro, T., Forget, F., Millour, E., Greybush, S.J.: Detection of detached dust layers in the Martian atmosphere from their thermal signature using assimilation. Geophys. Res. Lett. **41**(19), 6620–6626 (2014)
37. Arulampalam, M., Maskell, S., Gordon, N., Clapp, T.: A tutorial on particle filters for online nonlinear/non-Gaussian Bayesian tracking. IEEE Trans. Signal Process. **50**(2), 174–188 (2002)
38. Kong, A., Liu, J., Wong, W.: Sequential imputations and Bayesian missing data problems. J. Am. Stat. Assoc. **89**(425), 278–288 (1994)
39. Reid, H.: The Mechanics of the Earthquake, The California Earthquake of April 18, 1906, Report of the State Investigation Commission, vol. 2. Carnegie Institution of Washington, Washington (1910)

40. Daley, D.J., Vere-Jones, D.: An Introduction to the Theory of Point Processes, vol. I. Springer, New York (2003)
41. Werner M.J., Ide, K., Sornette, D.: Earthquake forecasting based on data assimilation: sequential Monte Carlo methods for renewal point processes. Nonlinear Process. Geophys. **18**, 49-70 (2011). www.nonlin-processes-geophys.net/18/49/2011/
42. Ogata, Y.: Seismicity analysis through point-process modeling: a review. Pure Appl. Geophys. **155**, 471–507 (1999)
43. Field, E.H.: A summary of previous working groups on california earthquake probabilities. Bull. Seismol. Soc. Am. **97**(4), 1033–1053 (2007)
44. Biasi, G., Weldon, R., Fumal, T., Seitz, G.: Paleoseismic event dating and the conditional probability of large earthquakes on the southern San Andreas fault, California. B. Seismol. Soc. Am. **92**, 2761–2781 (2002)
45. Marshall, A.: The use of multi-stage sampling schemes in Monte Carlo computations. In: Meyer, M. (ed.) Symposium on Monte Carlo Methods, pp. 123–140. Wiley, New York (1956)
46. Efron, B., Tibshirani, R.: An Introduction to the Bootstrap. Chapman and Hall/CRC, Boca Raton (1993). ISBN 0-412-04231-2
47. Liu, J.S., Chen, R.: Sequential monte carlo methods for dynamic systems. J. Am. Stat. Assoc. **93**(443), 1032–1044 (1998)
48. Mézard, M., Parisi, G., Virasoro, M.: Spin Glass Theory and Beyond. World Scientific Lecture Notes in Physics, vol. 9. Cambridge University Press, Cambridge (1987)

Appendix

A.1 Hadamard Product

For two real matrices A and B that are of the same dimensions.

1. If A and B are positive semidefinite, then is $A \circ B$,
2. If B is positive definite and if A is positive semidefinite with all its main diagonal entries positive, then $A \circ B$ is positive definite, Horn [1],
3. $A \circ (BC) \neq (A \circ B)C$
4. $A(B \circ C) \neq (A \circ B)C$

Points 3 and 4 will be needed for approximations in the inclusion of the Schur product in the EnKF and ETKF. These are now proved by writing the identities in a matrix index notation, where i is the row index, j is the column index and k is an index to be summed over.

$$
\begin{aligned}
[A \circ B]_{ij} &= A_{ij} B_{ij} \\
[(A \circ B)C]_{ij} &= \sum_{k=1}^{N} A_{ik} B_{ik} C_{kj} \\
[BC]_{ij} &= \sum_{k=1}^{N} B_{ik} C_{kj} \\
[A \circ (BC)]_{ij} &= \sum_{k=1}^{N} A_{ij} B_{ik} C_{kj}
\end{aligned}
\tag{A.1}
$$

R. Guzzi, *Data Assimilation: Mathematical Concepts and Instructive Examples*,
SpringerBriefs in Earth Sciences, DOI 10.1007/978-3-319-22410-7

Thus one concludes that $A \circ (BC) \neq (A \circ B)C$. Also

$$[A(B \circ C)]_{ij} = \sum_{k=1}^{N} A_{ik} B_{kj} C_{kj} \tag{A.2}$$

so it is now concluded that $A(B \circ C) \neq (A \circ B)C$.

A.2 Differential Calculus

Now we cover the various calculation methods used in previous chapters in order to provide a comprehensive and self-consistent framework of reference.

A.3 The Method of Characteristics

Let us consider the partial differential equations of the following form:

$$a(x, y)\frac{\partial u}{\partial x} + b(x, y)\frac{\partial u}{\partial y} + c(x, y)u = 0 \tag{A.3}$$

with the boundary conditions $u(x, 0) = f(x)$ where $u = u(x, y)$ is the unknown function we need to find, and the expressions $a(x, y)$, $b(x, y)$, $c(x, y)$ and $f(x)$ are given.

The objective of the method of characteristics , when it is applied to these equations, is to change coordinates from (x, y) to a new coordinate system (x_0, s) where the partial differential equations PDE becomes an ordinary differential equation (ODE) along certain curves in the plane $x - y$. These curves along which the solution of the PDE is reduced to ODE, are called characteristic curves or features. While the new variable s vary for along these characteristic curves, the new variable x_0 will be constant.

Let us now transform the PDE into ODE.

$$\frac{\partial u}{\partial s} = \frac{\partial u}{\partial x}\frac{\partial x}{\partial s} + \frac{\partial u}{\partial y}\frac{\partial y}{\partial s}. \tag{A.4}$$

If we select

$$\frac{\partial x}{\partial s} = a(x, y) \tag{A.5}$$

and

$$\frac{\partial y}{\partial s} = b(x, y) \tag{A.6}$$

we have:

$$\frac{\partial u}{\partial s} = a(x, y)\frac{\partial u}{\partial x} + b(x, y)\frac{\partial u}{\partial y} \tag{A.7}$$

and thus the PDE becomes the following ODE

$$\frac{du}{ds} + c(x, y)u = 0. \tag{A.8}$$

The relations (A.5) and (A.6) are the characteristic equations.

The strategy to be adopted to apply the method of characteristics is to:

1. solve quadratic equations (A.5) and (A.6);
2. integrate by placing constants $x(0) = x_0$ e $y(0) = 0$;
3. solve the ODE (A.8) with the initial conditions $u(x, 0) = f(x_0)$;
4. get a solution, solving it for s and x_0 in terms of x and y (using the results of step 1) and replace these values in $u(x_0, s)$ to get the solution of PDE $u(x, t)$.

A.4 Calculus of Variations

Calculus of variations has been developed by Leonhard Euler in 1744 to find the biggest or smaller values whose rate changed very quickly. In 1755 Lagrange wrote a letter to Leonhard Euler in which he described his method on the variations. Euler immediately adopted this new method called *Calculus of variations*. The great advantage of the calculus of variations is that we consider a system as a whole and the individual components of the system itself are not explicitly considered. In this way one allows to deal with a system without knowing in detail all the interactions between the various components of the system itself. The variational calculation determines the stationary points (extremes) integral expressions, known as functional.

We can say that a function $f(x, y)$ has a steady value at one point (x_0, y_0) if around this point, the rate of the function, in any direction, is zero. The stationary value concept is well described by the operator δ that was introduced by Lagrange. This operator is similar to the differential operator, but while this refers to a real infinitesimal displacement, the Lagrangian refers to an virtual infinitesimal displacement. The virtual nature of the Lagrangian operator arises from the fact that enables a shift around an esplorative point called change of position.

The operator δ behaves like the differential operator and vanishes at the endpoints of the curve defined by the function. Let us see how it works with a simple example.

Suppose there is a (x_0, y_0) around which we want to conduct a virtual displacement $\delta x, \delta y$. The first variation on function $f(x, y)$ is given by:

$$\delta f = \frac{\partial f}{\partial x}\delta x + \frac{\partial f}{\partial y}\delta y. \tag{A.9}$$

δx and δy can be written in terms of directional cosines α multiplied by a small parameter ϵ which tends to zero.

$$\begin{aligned} \delta x &= \epsilon \alpha_x \\ \delta y &= \epsilon \alpha_y. \end{aligned} \tag{A.10}$$

In this way the variation of the function, in a specific direction, is given by:

$$\frac{\delta f}{\epsilon} = \frac{\partial f}{\partial x}\alpha_x + \frac{\partial f}{\partial y}\alpha_y. \tag{A.11}$$

By definition, since (x_0, y_0) is a steady value, $\frac{\delta f}{\epsilon}$ must disappear for any virtual displacement with no regard to the direction of the shift that is independently on α_x and α_y. Then the condition that all partial derivatives disappear in a stationary point is a necessary and sufficient condition for the function f that has a steady value at that point.

$$\begin{aligned} \frac{\partial f}{\partial x} &= 0 \\ \frac{\partial f}{\partial y} &= 0. \end{aligned} \tag{A.12}$$

The fact that the value $f(x_0, y_0)$ is a stationary value of f is a necessary but not sufficient condition, to be an extremum of f. When the infinitesimal around $f(x_0, y_0)$ has $f > f(x_0, y_0)$ everywhere, then $f(x_0, y_0)$ is a local minimum. Vice versa if $f < f(x_0, y_0)$ then $f(x_0, y_0)$ is a local maximum. In this case the stationary value it is also an extremum. In other cases $f(x_0, y_0)$ is a stationary value but not an extreme. The second derivative allows us to understand if a stationary point is a maximum, minimum or nil. Of course we need to specify the domain where the stationary or extreme values can be found.

When the variational calculation has some constraints, for example such as:

$$g(x, y) = 0, \tag{A.13}$$

we use the Lagrange multipliers forming a new function $f_1 = f + \lambda g$ where λ is the Lagrange multiplier, a undetermined function. Then let us take the variation of the function $f(x, y)$ and the constraint

$$\delta f = \frac{\partial f}{\partial x}\delta x + \frac{\partial f}{\partial y}\delta y = 0$$
$$\delta g = \frac{\partial g}{\partial x}\delta x + \frac{\partial f}{\partial y}\delta y = 0. \tag{A.14}$$

Let us take the variation of f_1

$$\delta f_1 = \delta(f + \lambda g) = \delta\lambda + \lambda\delta g + g\delta\lambda = \delta f. \tag{A.15}$$

Then the conditions because there is a stationary point of f_1 subject to the constraint (A.13) is that:

$$\frac{\partial f}{\partial x} + \lambda\frac{\partial g}{\partial x} = 0$$
$$\frac{\partial f}{\partial y} + \lambda\frac{\partial g}{\partial y} = 0. \tag{A.16}$$

The two previous equations and the constraint (A.13) are used to find the stationary value. The Lagrange multiplier can be considered as a measure of the sensitivity of the value of a function f, in the stationary point, when it varies with the constraint given from relation (A.13).

In the case of N dimensions $x_1, \ldots, x_N$, for a stationary point of the function $f(x_1, \ldots, x_N)$ subject to constraints $g_1(x_1, \ldots, x_N) = 0, \ldots g_M(x_1, \ldots, x_N) = 0$, we have:

$$\frac{\partial}{\partial x_n}(f + \sum_{m=1}^{M}\lambda_m g_m) = 0 \qquad 1 \le n \le N. \tag{A.17}$$

A.5 The Solution for the Simplified Equation of Oceanographic Circulation

Let us look how to solves the (2.84) following Bannister [2]. The inclusion of errors in the field $f = f(x,t)$ and $W_f \int_0^L dx \int_0^T dt f(x,t)^2$ in the cost function, defines the inverse model as a *weak constraint* (Sasaki [3]).

Let us build the variations of $\mathcal{J}$ around the reference field $\hat{u}$ so we can see for what $u(x,t)$ $\mathcal{J}[u]$ is stationary.

$$\mathcal{J}[\hat{u} + \delta u] = \mathcal{J}[\hat{u}] + \delta\mathcal{J}\big|_{\hat{u}}\,. \tag{A.18}$$

Thus

$$\delta\mathcal{J}|_{\hat{u}} = \int_0^L dx \int_0^T dt \left.\frac{\partial J}{\partial u}\right|_{\hat{u}} \delta u + \mathcal{O}(\delta u^2) =$$

$$2W_i \int_0^L dx\{\hat{u}(x,0) - I(x)\}\delta u(x,0) + 2W_b \int_0^T dt\{\hat{u}(0,t) - B(t)\}\delta u(0,t)$$

$$+2w \sum_{i=1}^{M} \{\hat{u}(x_i,t_i) - u_i\}\delta u(x,t)\delta(x-x_i)\delta(t-t_i)$$

$$+2W_f \int_0^L dx \int_0^T dt \left\{\frac{\partial\hat{u}}{dt} + c\frac{\partial\hat{u}}{\partial x} - F\right\}\left\{\frac{\partial\delta u}{\partial t} + c\frac{\partial\delta u}{\partial x}\right\} + \mathcal{O}(\delta u^2). \quad \text{(A.19)}$$

Let us define

$$\hat{\mu}(x,t) = W_f(\frac{\partial\hat{u}}{dt} + c\frac{\partial\hat{u}}{\partial x} - F), \quad \text{(A.20)}$$

where $\hat{\mu}(x,T) = 0$ and $\hat{\mu}(L,t) = 0$. Integrating by parts

$$\int_a^b v\frac{du}{dx}dx = [uv]_a^b - \int_a^b u\frac{dv}{dx}dx, \quad \text{(A.21)}$$

a part of the integration

$$2\int_0^L dx \int_0^T dt \left\{\frac{\partial\hat{u}}{dt} + c\frac{\partial\hat{u}}{\partial x} - F\right\}\left\{\frac{\partial u}{\partial t} + c\frac{\partial u}{\partial x}\right\} = \quad \text{(A.22)}$$

$$2\int_0^L dx \int_0^T \hat{\mu}(x,t)\left\{\frac{\partial\delta u}{\partial t} + c\frac{\partial\delta u}{\partial x}\right\} dt \quad \text{(A.23)}$$

is written as

$$2\int_0^L dx \int_0^T \hat{\mu}(x,t)\left\{\frac{\partial\delta u}{\partial t} + c\frac{\partial\delta u}{\partial x}\right\} dt$$

$$= 2\int_0^L \hat{\mu}(x,0)\delta u(x,0)dx - 2\int_0^L dx \int_0^T \frac{\partial\hat{\mu}}{\partial t}\delta u(x,t)dt -$$

$$2\int_0^T c\hat{\mu}(0,t)\delta u(0,t)dt - 2\int_0^T dt \int_0^L c\frac{\partial\hat{\mu}}{\partial x}\delta u(x,t)dx. \quad \text{(A.24)}$$

Thus we have:

$$\delta\mathcal{J}|_{\hat{u}} = 2W_i \int_0^L dx\{\hat{u}(x,0) - I(x)\}\delta u(x,0)$$
$$+2W_b \int_0^T dt\{\hat{u}(0,t) - B(t)\}\delta u(0,t)$$
$$+2w\sum_{i=1}^{M}\{\hat{u}(x_i,t_i) - u_i\}\delta u(x,t)\delta(x-x_i)\delta(t-t_i)$$
$$-2\int_0^L \hat{\mu}(x,0)\delta u(x,0)dx - 2\int_0^L dx \int_0^T \frac{\partial\hat{\mu}}{\partial t}\delta u(x,t)dt$$
$$-2\int_0^T c\hat{\mu}(0,t)\delta u(0,t)dt - 2\int_0^T dt \int_0^L c\frac{\partial\hat{\mu}}{\partial x}\delta u(x,t)dx + \mathcal{O}(\delta u^2). \quad \text{(A.25)}$$

Posing the linear part to zero, using the Eq. (2.70) and the definition (A.20) we get the equations of Euler-Lagrange for a *weak constraints*, where:

1. the forward equation is:

$$\frac{\partial\hat{u}}{\partial t} + c\frac{\partial\hat{u}}{\partial x} - F = W_f^{-1}\hat{\mu}. \quad \text{(A.26)}$$

2. Its initial conditions and the boundary conditions are:

$$W_i\{\hat{u}(x,0) - I(x)\} - \hat{\mu}(x,0) = 0 \quad \text{(A.27)}$$
$$W_b\{\hat{u}(0,t) - B(t)\} - c\hat{\mu}(0,t) = 0. \quad \text{(A.28)}$$

3. The inverse is given by:

$$w\sum_{i=1}^{M}\{\hat{u}(x_i,t_i) - u_i\}\delta(x-x_i)\delta(t-t_i) - (\frac{\partial\hat{\mu}}{\partial t} + c\frac{\partial\hat{\mu}}{\partial x}) = 0. \quad \text{(A.29)}$$

4. Its initial conditions and at the boundary conditions are:

$$\hat{\mu}(x,T) = 0 \quad \text{(A.30)}$$
$$\hat{\mu}(L,t) = 0. \quad \text{(A.31)}$$

When we require that $u = u(x,t)$ satisfies exactly the Eq. (2.70), then we need to find a minimum for the functional:

$$f[u] = W_i \int_0^L \{u(x,0) - I(x)\}^2 dx + \tag{A.32}$$

$$W_b \int_0^T \{u(0,t) - B(t)\}^2 dt + w \sum_{m=1}^{M} \{u(x_i,t_i)u_i\}^2.$$

The query is: which $u(x,t)$ makes $f[u]$ stationary, subject to the following model constraint?

$$g(x,t) = \frac{\partial u}{\partial t} + c\frac{\partial u}{\partial x} - F = 0. \tag{A.33}$$

This question is answered by adding to the functional $f[u]$ the *strong constraint* obtained using a Lagrange multiplier $\lambda = \lambda(x,t)$ obtaining:

$$\mathcal{J}[u,\lambda] = f[u] + 2\int_0^L dx \int_0^T \lambda(x,t)g(x,t)dt. \tag{A.34}$$

Note that u and λ can vary independently. The variation of $\mathcal{J}$ around the reference field $\hat{u}$ and $\hat{\lambda}$ is given by:

$$\mathcal{J}[\hat{u}+\delta u, \hat{\lambda}+\delta\lambda] = \mathcal{J}[\hat{u},\hat{\lambda}] + \delta\mathcal{J}\Big|_{\hat{u},\hat{\lambda}} \tag{A.35}$$

where

$$\delta\mathcal{J}|_{\hat{u},\hat{\lambda}} = \int_0^L dx \int_0^T dt \frac{\partial \mathcal{J}}{\partial u}\Big|_{\hat{u},\hat{\lambda}} \delta u + \int_0^L dx \int_0^T dt \frac{\partial \mathcal{J}}{\partial \lambda}\Big|_{\hat{u},\hat{\lambda}} \delta\lambda + \mathcal{O}(\delta u^2, \delta\lambda^2, \delta u\delta\lambda). \tag{A.36}$$

Thus

$$\begin{aligned}
\delta\mathcal{J}|_{\hat{u},\hat{\lambda}} = 2W_i \int_0^L & dx\{\hat{u}(x,0) - I(x)\}\delta u(x,0) \\
+2W_b \int_0^T & dt\{\hat{u}(0,t) - B(t)\}\delta u(0,t) \\
+2w \sum_{i=1}^{M} & \{\hat{u}(x_i,t_i) - u_i\}\delta u(x_i,t_i) \\
+2\int_0^L dx \int_0^T & dt\hat{\lambda}(x,t)\left(\frac{\partial \delta u}{\partial t} + c\frac{\partial \delta u}{\partial x}\right) \\
+2\int_0^L dx \int_0^T & \delta\lambda(x,t)\left(\frac{\partial \hat{u}}{\partial t} + c\frac{\partial \hat{u}}{\partial x} - F\right)dt \\
& +\mathcal{O}(\delta u^2, \delta\lambda^2, \delta u\delta\lambda),
\end{aligned} \tag{A.37}$$

with the boundary conditions $\hat{\lambda}$: $\hat{\lambda}(x, T) = 0$ and $\hat{\lambda}(L, t) = 0$.

Integrating by parts (A.37), that is:

$$2\int_0^L dx \int_0^T dt\hat{\lambda}(x,t)\left(\frac{\partial \delta u}{\partial t} + c\frac{\partial \delta u}{\partial x}\right) \tag{A.38}$$

with

$$\begin{aligned} 2\int_0^L dx \int_0^T dt\hat{\lambda}(x,t)\left(\frac{\partial \delta u}{\partial t} + c\frac{\partial \delta u}{\partial x}\right) &= \\ \delta u(x,T)\hat{\lambda}(x,T) - \delta u(x,0)\hat{\lambda}(x,0) - \int_0^T dt\delta u\frac{\partial \hat{\lambda}}{\partial t} & \\ +c\delta u(L,t)\hat{\lambda}(L,t) - c\delta u(0,t)\hat{\lambda}(0,t) - \int_0^L dx c\delta u\frac{\partial \hat{\lambda}}{\partial x}. & \end{aligned} \tag{A.39}$$

Let us note the observation term is

$$\{\hat{u}(x_i,t_i) - u_i\}\delta u(x_i,t_i) = \int_0^L dx \int_0^T \{\hat{u}(x_i,t_i) - u_i\}\delta u(x,t)\delta(x - x_i)\delta(t - t_i). \tag{A.40}$$

With these solutions and the condition on $\hat{\lambda}$ we obtain the complete result

$$\begin{aligned} \delta\mathcal{J}|_{\hat{u},\hat{\lambda}} &= 2W_i\int_0^L dx\{\hat{u}(x,0) - I(x)\}\delta u(x,0) + 2W_b\int_0^T dt\{\hat{u}(0,t) - B(t)\}\delta u(0,t) \\ &+ 2w\sum_{i=1}^{M}\{\hat{u}(x_i,t_i) - u_i\}\delta u(x,t)\delta(x - x_i)\delta(t - t_i) \\ &- 2\int_0^L dx\delta u(x,0)\hat{\lambda}(x,0) - 2\int_0^L dx\int_0^T dt\frac{\partial \hat{\lambda}}{\partial t}\delta u(x,t) \\ &- 2\int_0^T c\delta u(0,t)\hat{\lambda}(0,t)dt - 2\int_0^T dt\int_0^L c\frac{\partial \hat{\lambda}}{\partial x}\delta u(x,t)dx \\ &+ 2\int_0^L dx\int_0^T \delta\lambda(x,t)\left(\frac{\partial \hat{u}}{\partial t} + c\frac{\partial \hat{u}}{\partial x} - F\right)dt + \mathcal{O}(\delta u^2, \delta\lambda^2, \delta u\delta\lambda). \end{aligned} \tag{A.41}$$

Posing the linear part to zero and using the boundary conditions of $\hat{\lambda}$ we get the Euler-Lagrange equation for a *strong constraint*,

1. where the forward equation is:

$$\frac{\partial \hat{u}}{\partial t} + c\frac{\partial \hat{u}}{\partial x} - F = 0 \tag{A.42}$$

2. the boundary conditions are:

$$W_i\{\hat{u}(x,0) - I(x)\} - \hat{\lambda}(x,0) = 0 \quad (A.43)$$

$$W_b\{\hat{u}(0,t) - B(x)\} - c\hat{\lambda}(0,t) = 0. \quad (A.44)$$

3. The inverse equation is:

$$w\sum_{i=1}^{M}\{\hat{u}(x_i,t_i) - u_i\}\delta(x - x_i)\delta(t - t_i) - (\frac{\partial\hat{\lambda}}{\partial t} + c\frac{\partial\hat{\lambda}}{\partial x}) = 0, \quad (A.45)$$

4. with the boundary conditions

$$\hat{\lambda}(x,T) = 0 \quad (A.46)$$

$$\hat{\lambda}(L,t) = 0 \quad (A.47)$$

References

1. Horn, R.: The Hadamard product. Matrix theory and applications. In: Johnson, C.R. (ed.) American Mathematical Society, Proceedings of Symposia in Applied Mathematics, vol. 40, pp. 87–169 (1990)
2. Bannister, R.: Various papers on data assimilation. http://www.met.rdg.ac.uk/~ross/DARC/DataAssim.html (2012)
3. Sasaki, Y.: Some basic formalisms in numerical variational analysis. Mon. Weather Rev. **98**, 875–883 (1970)

Index

R. Guzzi, *Data Assimilation: Mathematical Concepts and Instructive Examples*,
SpringerBriefs in Earth Sciences, DOI 10.1007/978-3-319-22410-7

MIX
Papier aus verantwortungsvollen Quellen
Paper from responsible sources
FSC® C105338

If you have any concerns about our products,
you can contact us on
ProductSafety@springernature.com

In case Publisher is established outside the EU,
the EU authorized representative is:
Springer Nature Customer Service Center GmbH
Europaplatz 3, 69115 Heidelberg, Germany

Printed by Libri Plureos GmbH
in Hamburg, Germany